CARDINAL POINTS:
NORTH

SUSAN SCHROEDER

authorHOUSE®

AuthorHouse™
1663 Liberty Drive, Suite 200
Bloomington, IN 47403
www.authorhouse.com
Phone: 1-800-839-8640

First published by AuthorHouse 2/21/2008

ISBN: 978-1-4343-4633-9 (sc)

Printed in the United States of America
Bloomington, Indiana

This book is printed on acid-free paper.

For Leonard,
We miss you so much.

For Arlene, who has had the strength all of these years to
sustain a loving family, full of Christian values.

Thank you for all your help,

Priscilla and Bosco Westrich

Howard, my love,

And a special thank you to Jane Zeiser, for all of your help
and guidance.

CONTENTS

Chapter One
The Water Balloon Victory   1

Chapter Two
All Wet   11

Chapter Three
Dark and Creepy   15

Chapter Four
The First Day   25

Chapter Five
What Goes Around Comes Around   33

Chapter Six
All Messed Up   41

Chapter Seven
Grandma's Treasure   49

Chapter Eight
Pete Exceeds All Expectation   57

Chapter Nine
I Pledge Allegiance to the Friendship...   69

Chapter Ten
A Strange Phone Call   75

Chapter Eleven
Davy's Turn   77

Chapter Twelve
The Midnight Dance   85

Chapter Thirteen
Billy's Turn   95

Chapter Fourteen
All That Glitters   103

Chapter Fifteen
The Odd Figure-8   107

Chapter Sixteen
Tragedy Falls   111

Chapter Seventeen
Morganite   117

Chapter Eighteen
Numbers and Letters   123

Chapter Nineteen
The Watcher in the Woods   131

Chapter Twenty
What's the Buzz?   139

Chapter Twenty - One
The 'Equation of Time'   149

Chapter Twenty - Two
Someone in the Shadows   153

Chapter Twenty - Three
The Sacred Book   157

Chapter Twenty - Four
A Sweet Reunion   161

Chapter Twenty - Five
Coming Home   165

Chapter Twenty - Six
Digging-Up the Past   173

Chapter Twenty - Seven
The Secret of the Headstone   181

Chapter Twenty - Eight
S H E   187

Chapter Twenty - Nine
Old News   199

Chapter Thirty
Tourmaline, Garnet & Ametrine   203

Chapter Thirty - One
Game Set   207

Chapter Thirty - Two
A Bad Rash   211

Chapter Thirty - Three
From the Rubble   227

Chapter Thirty - Four
A Precious Ending   229

CHAPTER ONE
THE WATER BALLOON VICTORY

He acted alone. On his way to meet Charlie McMullin he saw that Mrs. Elders' peach tree was full and ripe and he couldn't resist. Billy crawled on his belly under the wooden fence. The grass was still damp from the morning dew, and he drug himself into the old lady's yard. When he was in the yard, he slowly stood up. The morning sun was warm on his face and shoulders, and he knew if Mrs. Elders' was in her kitchen she would see him out of her window. He moved slowly over to the tree where the peaches seemed to tease him. He reached up and put his fingers on the largest ripe, juicy peach he could find, and then slowly moved it towards his mouth. Every move deliberate so not to draw attention to himself. The peach juice exploded in his mouth, and he chewed around the seed. He tossed the seed on the ground. He had to have more.

Pulling his shirt up, he began picking as fast as he could, dropping the peaches into the pocked he formed with his shirt. He looked around with a guilty glance to see if anyone saw him in the yard. He had his shirt almost full when he heard the voices coming from the house. The back door opened slowly, and there he was, caught. At first, no one seemed to see him, so he ducked behind the bulk of the tree trunk and stood as still as he could. A bee started buzzing around his head, but still he stood completely still. Billy Martin wasn't considered the toughest kid in the seventh grade for nothing. The bee landed on his hand and began to move up his arm. It was hard to keep an eye on the bee and the people moving towards him at the same time.

The bee disappeared under his sleeve, and that was all Billy could stand. He switched hands on his shirt, careful not to spill any of the peaches, and with the other, he reached into his sleeve to retrieve the bee. When he did, he brushed the bee and it stung his arm. Billy let out a scream, and dropped his peaches on the ground.

The other people in the yard looked at the direction of the sound, and saw Billy Martin doing what looked to be a jig behind the peach tree. "Just what are you doing back there young man?" asked Mrs. Elders. Billy looked up and saw that Pete James and Davy Baker were standing next to Mrs. Elders with bucket in their hands, and looking at him with surprise.

Billy glared at the boys with daggers in his eyes. That's all he needed was for Pete and Davy to start smarting off. These two were his nemesis in the seventh grade. If Pete, one of the smartest, fastest, and all around goodie-two-shoes, said anything at all, he would pound his face into the grass. But it wasn't Pete that spoke, it was the wily little Davy Baker that always started the trouble. Davy was small, wiry, and seldom thought before he spoke. "Hey look, it's that big pig Billy Martin stealing your peaches."

Billy couldn't take the smart crack from such a small, stupid jerk like Davy. He forgot about the bee sting on his arm, and ran around the tree as fast as he could. Davy and Pete dropped their buckets and ran towards the big kid coming at them. Billy tried to hit both boys straight on with his shoulder hoping to knock them off their feet, but the two boys suddenly moved apart and Billy hit the ground hard.

He looked up and saw both boys had drawn their hands into tight fists and were ready to pound on him. Billy quickly rolled over onto his side and sprang to his feet. There was no way he was going to let these two goodie-goodies pound on him, not when he was the biggest, and toughest kid around. Once it got out that these kids beat on him, his reputation would be ruined. He drew in a breath and made a terrible sound with his mouth and spit towards both boys. Then he ran towards the fence, throwing himself on his belly, and crawling as fast as he could out of the yard.

Stunned to be hit in the face with spit, the boys stood watching for a minute, and then sprang after the bully. Davy grabbed at Billy's foot, but Billy kicked him off with one jerk. He was under the fence and gone before the boys could get to him. "I'll catch you and pound you like a piece of chicken, like the chicken you are, Billy you jerk," yelled Davy. "You would have to catch me first, you little baby," yelled Billy from over the fence. Davy put a foot on Mrs. Elders' fence and was ready to hop over when he heard her voice, "Davy, please get off the fence." Davy turned and saw that she had picked up the buckets and was handing them back to Pete and himself. Davy jumped off the fence and took the bucket from Mrs. Elders.

"Boys, forget about him, he's gone and I need you to pick these peaches before the birds get to them. Remember I'm going to pay you $2.00 a bucket." She smiled warmly and rubbed Davy's head. Davy stepped back and shook his head, embarrassed to be touched like a little kid. "Thank you, we'll get started," said the other boy.

Mrs. Elders looked at the two boys and turned to go inside. When she was in the house, Davy turned to Pete and gave him a long stare. "We could have taken that big pig, you and me, why didn't you back me up?" Pete, the tawny haired, brown eyed boy with perfectly straight teeth smiled an even smile and said. "We're getting paid to pick the peaches, and we can eat as many as we want, why bother with a bully like Billy when we are coming out ahead?" Davy turned and looked at the tree, it was loaded with ripe peaches. He knew he would eat as many as he put in his bucket, and he shook his head and agreed with Pete.

Davy, with his dark curly hair and big blue eyes, was an impulsive sort of guy. Until Pete had moved to town two years ago, he really didn't have anyone to hang out with. He was smaller than the other guys, and Billy was always picking on him. Now that he had Pete, he seemed to get more respect. Pete was the most liked kid in the whole seventh grade. The teachers liked him, the girls really liked him, and most of the guys liked him. He was the coolest kid Davy ever met, and Davy stuck to Pete like burs stick to your socks.

Davy gave a snort and said, "I bet that big pig hurt his shoulder when he hit the ground, he must weigh about 150 pounds, he's such a big chicken, he didn't stick around because he knows the two of us could pound him into the mud....where do you think he's going? I bet he's going to run off and tell his buddy Charlie McMeat-head that he beat us up. What a big pig......" and off Davy went into his endless chatter.

Pete drew in a long breath and let out a heavy sigh. He heard his buddy chatting away and drifted off into his own thoughts. Pete had a hard time concentrating on anything in his life. All of his free thinking was consumed by worry for his father. Pete was proud of his father, but at the same time was angry that he could leave them and pursue his career in such a dangerous place. Pete's mom and two little sisters had moved back to the small town of Preston two years ago. They lived with Pete's grandparents in a house that was more than 200 years old. It was a large estate, and the house had plenty of room for everyone, but it was so old. There was a dirt basement and it smelled like turnips, the wallpaper was yellow, flowery and peeling off the walls, and there were rooms every where. It wasn't like the house they lived in when they lived in the city. The city house was modern and new, and this house was old and falling apart. Grand-dad was a really good guy and spent a lot of time with Pete, but he couldn't look after the house anymore.

Pete's father was a news correspondent for a large television station. Pete's father had agreed to take a job in Iraq to report the news during the war. It was only supposed to be for a couple of months, and then it turned into a year, and now two. Pete's mother had little say as to what her husband wanted to do, and so they agreed that Pete's mom and sisters would come to live here until he got back. It has been more than two years now, and with every passing day, Pete worried that he would never see his father again. He often heard his mother crying in her bed, or moaning aloud overcome with grief while watching the evening news. Every time Pete had a moment to himself, he was left wondering what issues his father was dealing with. He wished his father would return home and never leave them again.

"Well, what's it going to be?" Davy was looking at him, and he must have repeated himself more than once, because he was standing there staring at Pete. "Earth to Petie, are you listening?" Pete looked directly at Davy and said, "well, yeah, I'm just getting the peach juice off my fingers," and he took his hand and wiped it on his shirt. "How long did you want to do this? I'm already full with my first bucket." Changing the subject from whatever Davy was saying, Pete strode into the house and came back with another bucket.

"How did you fill that bucket so fast?" Davy asked. "I'm picking and not talking," replied Pete. Davy looked down into his bucket and saw that he had barely half filled his bucket. He moved to another limb of the tree and started to pick peaches with a fierce urgency, that if he didn't fill his bucket soon, he would loose his chance to earn the money.

The picking of the peaches became the sole effort from both boys as the silence surrounded them. Pete heard the tufted-tit mouse sing, "Peter, Peter, Peter" off in the distance. Needless to say, this was Pete's favorite bird. His father was a birder, and Pete could recognize some of the resident birds by song. He heard the black-capped chickadee sing, "Chickadee-dee-dee" and he smiled to think that probably no other kid in the seventh grade could identify a single bird by sight or song. Pete knew the song of the eastern fly-catcher, he said, "Pewee," or "Feebee." The cardinal said, "Birdie, birdie, birdie," and the barred owl said, "Who cooks for you, who cooks for you, who cooks for you all?" The other kids were impressed that he would name some of the birds just from the songs they sung. This made Pete appear smart and he liked to think the other kids thought he was smart.

At just about noon, the boys handed in their last buckets and were paid $6.00 each. Davy stuck the money deep into his pocket and ran out the front door letting it slam behind him. Pete said thank you and made a little more small talk, and then politely closed the door so it wouldn't slam. He strode down the sidewalk toward Davy who was waiting just beyond the driveway.

"What do you want to do with your money? We could go to the movies and see the new Superman movie, you wanna do that?" Davy asked. "Or we could go get those new b-giant squirt guns and blast Billy and Charlie when they're coming out of Billy's house." Billy lived practically next door to Davy. "Or, we could go the Dairy Queen and have double cheese burgers and brownie delights."

Davy often asked many questions without waiting for a response. Pete scratched his head and looked around. "I don't know," and really he didn't. All of the possibilities sounded good, and he was thinking about which one he would enjoy the most.

It was the sound of dogs barking that made up his mind. From behind them, the sound grew louder and made them turn their heads to see that Billy Martin and Charlie McMullin were leaning over the fence at Mrs. Elders' backyard looking for something. The two boys moved cautiously behind the hedges that lined Mrs. Elders' front yard. From where they stood, the two boys in the back yard could not see them, but Davy and Pete could see everything that was going on.

Charlie and Billy were looking for them. Billy must have told Charlie, and now both boys were in pursuit of Davy and Pete. The idea must have come to them at the same time, because both boys said "water guns" under their breath at exactly the same time.

Squatting as they turned, the boys quietly headed off towards the Quick Shop three blocks away. They kept their heads ducked down as they ran swiftly down the street. When they reached the Quick Shop, they strolled in standing straight up, and moved to the isles with the kiddy stuff. There were only three water guns left, two were medium size and one was giant size. It had a special long tube for ice cubes so you could fill up the gun with ice water.

Davy grabbed the large one leaving the smaller ones for Pete. Davy looked around and seemed concerned that he was going to get the big gun and Pete would have to take a smaller one. Pete shook his head, as if to read his friends mind, "It's okay, really, I'll take the smaller one, and ..." he grabbed a bag of balloons and said, "we'll nail them with balloons too". Davy gave a big smile and they practically ran to the front of the store to purchase their ammo.

They ran out of the store to the back of the building and down the alley to Davy's backyard. His mom was inside with his little brother, and didn't seem to mind the boys playing around the house. Davy ran to the shed and pulled out two buckets and a small cooler. "We could fill up the balloons and stash them around the front yard." Davy's eyes were huge as he explained his plans for the attack. Pete could see that years of taking Billy's abuse were coming to a head today as he carefully planned his revenge.

The boys worked on filling the balloons in quiet agreement. Davy filled them while Pete tied and placed them in the buckets and cooler. When they were done, they filled the guns. Davy ran inside for a cup of ice and poured it into the large container for the ice water. He poured the ice in his gun; nothing meaner than an ice water assault.

Together, they lifted the buckets of balloons, heavy with water, and walked them out to the front yard. They looked around every corner to make sure they weren't being watched by the other boys. Davy placed his bucket next to a big oak tree where the trunk of the tree hid the bucket from the view of anyone on the street or the house next door. Pete put his bucket behind the wheels of Davy's mother's car that was parked in the drive way. They went back and grabbed the cooler and put it on the side of the house, so if they had to retreat, they had a second stash of balloons. Then they grabbed their guns and took their post; Davy behind the bushes that bordered the driveway, and Pete behind the car.

Quiet anticipation filled Pete. He wasn't afraid of Charlie, he was his size but looked more frail. Billy was a head taller and probably 40 pounds heavier. Billy could be rough and mean at the drop of a jellybean. Billy bullied all the boys at school, and called the girls names. Charlie really was Billy's only friend.

Charlie arrived in town at about the same time as Pete. Given different circumstances, the two might have become friends, but Pete decided that Charlie was a spoiled brat living in a huge house with fancy cars. It appeared that Charlie got everything he wanted. Charlie was smart and a good athlete with blond hair and green eyes. He was clearly the best looking kid in the seventh grade. The girls liked Charlie because he was so handsome, but they did not like Billy, and so Charlie was "bad" by association. Most of the kids stayed away from both of them, but today was the day that Pete and Davy planned to even the score.

It was only a 10 minute wait for Pete and Davy until the other two boys showed up. They came walking down the street quite loudly. Billy was looking around, trying to see if he could find Davy. Davy had told Pete that many times in the past Billy would chase him down, push his face in the ground and punch him in the sides. Then he would drag him off the ground and throw him in the sticker bushes and run off laughing. Davy was so tired of being bullied, that he would stay in his house and look out the windows most of the day. That was before Pete came to town. Now Davy took on an air of strength and started to fight back.

Pete put his finger to his lips and then flattened the palm of his hand in a signal that meant, wait. When the boys reached the front of the next house, about 3 feet away, Davy jumped up and started squirting them with his gun. His gun had the feature that gave a continuous spray without pumping. Pete grabbed a handful of balloons and started pelting them at the two boys. The first balloon bounced off Billy and he yelled, "Hey! I'm going to kill you for that." Pete's second balloon broke on his side and water splashed over both of them. He continued throwing balloons at a record pace, and then Davy started too.

At first, the other two boys just moved to avoid being hit. This lasted for a couple minutes, and then Charlie started to run in on Pete. That's when Pete let him have it with the squirt gun. He sprayed him up and down and then right in the face. Charlie took his hands and covered his face only to become a target for the water balloons again. A balloon broke on his legs, his waist and then on his head.

Billy was screaming and yelling, but Davy was holding his own. He squirted him mainly in the face, and when he covered his eyes, Davy would nail him with water balloons. "How you holding out?" yelled Pete to Davy. "I'm ready for plan B," yelled Davy back. "Okay, grab your last ones and let's move," yelled Pete.

Both boys quickly grabbed the last of the balloons and their guns and ran to the side of the house. They picked up the cooler and scooted towards the back, but not around the corner. "You look out front and I'll look out back," yelled Pete. They went to their positions, and sure enough Billy was coming to the front, and Charlie had come around the back.

Round two had begun as the boys were getting pelted with balloons for the second time. This time Pete took a handful of balloons and tucked them into his shirt like a pouch and was running after Charlie, hitting him in the back and head. Charlie had no recourse but to run off up the street. Pete turned and went around the house the opposite way. He came up behind Billy and hit him hard from the back breaking balloons on his big butt and shoulders. Billy turned and saw balloons coming from both sides and ran off up the street behind Charlie. Pete ran after him throwing as many balloons as he had tucked in his shirt.

Pete started jumping up and down in the street. "Yeah!" he screamed. He ran back to where Davy was standing, "That was so great, Yeah!" he screamed again. Davy sank back against the house. "Yeah!" he said exhausted. "I will never forget this day," he said with a heavy sigh, and sat heavily on the ground. He looked up and smiled thankfully at Pete. "I never thought I could get the best of that jerk," and he sucked in a huge breath of air. It was almost as if years of torture had been repaid in a single afternoon.

Davy's head sunk into his chest as he drew in deep breath. It was as if he had overcome a huge crisis and was recovering from the experience. Pete watched him in silence. He understood that this was big for Davy.

"Davy, we have to move out of here, they may come back and try to get us"said Pete. "What do you say we go to my house and see what Pops is up to?" Pops is what Pete started calling his granddad. It's what his father called him, and he liked the way it sounded.

Davy looked up with watery eyes, "Yeah, let's put the buckets back and get out of here." Pete offered Davy his hand and pulled him off the ground. He patted his back and said, "We done good, buddy."

CHAPTER TWO

ALL WET

Water dripping from their eyes, Charlie and Billy walked down the road. "Little jerks, who do they think they are?" moaned Billy as he limped down the street. "I'm going to pummel them both in the ground next chance I get." Charlie looked over at his large, overgrown friend and stopped walking. He looked into the muddy green eyes of Billy Martin and said, "Are you kidding me, they planned that out, they waited for us, and executed an organized attack. All you do is bully everyone, but they sacked us good."

Billy looked at Charlie surprised, "Are you saying you're glad they got us?"

"No you big meat head, all I'm saying is that they thought it out, it wasn't just random."

"So what now, we just let them get away with it? Just let them think they got us? I'm the toughest kid in class and they're going to tell everyone what they did." Billy was mortified by the thought that the scrawny Davy Baker and his nerdy friend Pete Jones were going to tell everyone that they creamed Billy Martin.

Billy put his thick hands on his hips and stomped his feet, "I am not going to be made a fool by those two idiots."

Charlie shook his head and said, "Billy, we have to be smarter than them is all I'm saying, and that was a well thought out plan. That Pete is one smart kid, and if we are going to retaliate, it has to be good."

Billy was mad. His face was large and swollen. Even his small ears were red. Charlie on the other hand thought the attack was clever and fun. He never had a chance to experience a real water balloon fight, even though he was on the losing end, it was great.

Charlie walked next to Billy in silence and thought about how they had executed the plan. It was simple, yet effective. He thought about Pete and Davy and how they worked together at the right moments to pull it off. He looked at Billy and doubted if Billy could pull off something so genius and exact.

Billy was a bit of a fool, with a fowl temper, but he was his friend. Since moving to this small town, Charlie really had very little choice of who he was going to hang out with. His mother remarried and this is where the man she married wanted to be. His stepfather was friends with Billy's father, and so they were thrown together as friends. Certainly, Billy would not have been his choice, but he was trying to appease his stepfather.

Charlie turned around slightly to see if he could see either of the two boys that had besieged them with the attack. No one was there. They were probably off celebrating their well won victory, or hiding from Billy.

Charlie would have liked to be part of such a victory. He was a smart kid, and he resented being left with the big jerk. If he had a choice, he would have made friends with Pete and Davy, but his stepfather was insistent that he be friends with Billy.

Charlie stopped and looked at Billy, "Hey, what do you say we offer peace and ask if they want to play capture the flag?"

Billy scowled at Charlie, "Are you kidding me? Those two bozos are going to tell everyone that they got us, you think that's going to be cool? I'm going to look like a laughing cow."

Charlie thought about this for a minute. If Davy, who could not stop from talking it up, and Pete were to tell anyone in class what just happened, Billy would indeed look foolish. So would he, but he really didn't care as much as Billy. It was all just a part of being in the seventh grade.

Charlie looked at Billy and a smile crossed his face. "Well then my good friend, we will just have to plan our own revenge."

This made Billy smile too. Billy shook his head and started walking, "I am going to kick their butts, and they are going to be sooooo dead."

Charlie pulled him back and said, "Not the way to fame, we don't want to hurt them, just make them look like fools, like the way they did us."

Billy thought about this a moment. Pain definitely would make him feel better. "I guess that's the way to go," and a mean smile covered his face he and said under his breath, "but if I hurt them on accident, it's not my fault."

CHAPTER THREE

DARK AND CREEPY

All was quiet as it should be. Davy went about his business for days without worrying about being pushed into the sticker bushes, or being called names. Before he went outside to bring in the trash cans, he would first look to see if he was going to be assaulted by Billy. He was feeling triumphant, but very cautious at the same time.

School would start again in only two weeks as summer was drawing to an end, and he knew that Billy would get him back soon. He could only imagine how badly he was going to get it.

Ever since Davy was a little boy, the Martins lived two doors down. Billy began calling him names in preschool, and then would take his milk money in kindergarten. By the time they were in the second grade, Billy had beat him up three time, and once he knocked out his tooth. It wasn't permanent, but still it hurt. Davy lived in constant fear of the bully, but since the water balloon incident, he was beginning to feel so much better.

Davy lived with both of his parents in a relatively small house. His father worked in town at the power company, and his mother stayed at home and took care of him and his younger brother. His brother was four, but something was not right with him. He looked okay, but he didn't want to be held, or touched, and he didn't seem to notice if you were in the room or not. His mother spent most of her time with him, and Davy was often sad that he didn't seem to warrant attention from either his mother, or his little brother.

His mother used to drag him along to doctor after doctor to see what was wrong with his brother, and after what seemed to be an eternity, they came up with the diagnosis that Davy couldn't pronounce or explain. They said he had autism or something, and that he was just going to be like that.

During the day Davy kept to himself. Before Pete came to town, he had created his own private world. He made up stories and wrote them in journals. He wrote of large scaly dragons eating knights, he wrote of jungles filled with wild animals that chased and ate villagers, and he wrote of little mighty boys who would fight and beat up bullies. Davy had a world of adventures in his head that went on and on. He filled three journals by the time he was ten, and was working on his newest adventure in space with real ugly space monsters.

Despite the fact that Davy was shorter and skinnier than most of the kids in his class, he was quick and agile. His hair was curly and wild, his ears seemed to stick out and his teeth were coming in large with spaces between them. His face was splattered with red freckles, and because of his size and looks, most other kids didn't give him the time of day.

Although Davy was smaller, it didn't stop him from being the noisiest kid around. Davy seemed to always be making noise. Even in his sleep he was noisy. Pete didn't seem to mind all the noise, in fact it was a great distraction to his own confusing world.

Davy was the main source of Billy Martin's aggression. It seemed that ever since Davy was little, Billy would tease him and call him names. As they got older, Billy was more aggressive and would physically torment him. Davy avoided Billy constantly, and would often hide when he saw Billy coming down the street.

It was the middle of August and Pete was sitting on the dusty floor in Davy's room. He was thumbing through the journal of a boy who was on safari in the Amazon jungle and it occurred to him that what they needed was an adventure. Davy was busy writing in his new journal, he was biting on his lower lip making a clucking sound with his tongue. "Davy, what do you say we have our own adventure, today?"

Davy looked up in mid-scribble and said, "Adventure? Like going down to the lake and catching frogs and putting them in Mr. Jenkerson's pond?" Pete rubbed his chin, the thought of Mr. Jenkerson chasing large frogs out of his little backyard pond would be fun, but no, he wanted to do something bigger.

"I was kind of thinking about finding hidden treasure, or a pirate's map, or something big?" Pete said with wide eyes. Davy sat right up on his bed and put down the journal and pen. Pete had gotten Davy's attention. "You know my grandparents live in that old house, and I bet there is something good hidden somewhere in it."

Davy's eyes were wide with intrigue. Instantly Davy began to make a low "hmmmmmm" sound. Pete could see that Davy's mind was a whirl. What was going on inside his little peanut brain?

"We could start in the basement, its really big and kind of creepy" said Pete. "Do you have a flashlight?" Davy reached under his bed and pulled out a long, heavy black flashlight. "Ready to go," he said. They pulled on their shoes and crept quietly out of Davy's house. As they walked down the street, Davy started to talk non stop about the adventure they were to embark upon. Pete was trying to remember where Pops kept the flashlight.

When they got to the house, they went in the back door and into the mud room. Pete was sure he could find a flashlight in one of the cabinets there. The two boys were furiously looking in the cabinets when Pops came in to see what the ruckus was all about.

"You want to tell me what you lost?" asked Pops. The boys looked at each other and then Pete said, "Pops, can I borrow your flashlight?"

"What do you need a flashlight for when the sun's out?"

Pete, unsure that he should tell Pops what they were up to, looked at Davy who immediately looked at his shoes. "You boys going somewhere dark?" said Pops. "We want to look around the basement for stuff?" said Pete. Pops sat down heavy in the rocking chair in the corner of the room and began to rub his chin. "The only stuff you're gonna find in the basement is old and smelly. You sure you want to go down there? I remember when I was a young man and I wanted to look around in the basement, it didn't take me long to realize that what was down there wasn't anything I wanted to have."

"Well, still I want to see what's down there," said Pete in a voice that didn't sound so sure.

"Well then you will need a flashlight," grumbled Pops, and he stood up. He reached into the cabinet over the sink he pulled out a fat, red flashlight. "Bring this back when you're done," he commanded as he handed it to Pete. "Excellent," Pete said and took the flashlight.

"Oh, and boys, don't tell your Grandma I said you could go down there, she may not like you being in the basement." The boys looked at each other, and left the room rather quietly.

The boys went to the back of the house where the back stairs to the basement was located. The house was so large that it had two sets of stairs to the basement. The door was old and the paint was peeling. When Pete opened the door, it let out a low, creaking noise. Pete slid inside the small opening and Davy followed. The rancid smell of turnips, and dirt hit them as soon as they started down the steps. White light streamed in through the dirty windows and illuminated long wisps of cobwebs hanging from the rafters in the ceiling. Pete ducked under a long one, and Davy, who wasn't looking where he was going, got a face full as he walked behind Pete. "Oh," Davy said with a growl. "Disgusting." Pete turned to see Davy pulling a long cobweb off his face and shoulders. "I think you need to look where you're going," Pete whispered.

Neither boy moved. They looked at each other like they both really wanted to leave. Pete spoke first, "You know this is going to be creepy, but if we want to find treasure, we have to keep going." Davy only shook his head and then started making growling noises.

Pete turned on his flashlight, although he could see from the sunlight coming from the window. "Are there any lights down here?" Davy asked. Pete turned his flashlight to the ceiling and he saw bare light bulbs with long strings coming down. He reached up and pulled the string. A bright, yellow light flooded the basement. He walked slowly over to another string and pulled it once, nothing happened, he pulled it again, nothing happened. "Well, not all the lights seem to work," Pete said, and looked around for more lights. Davy was on the other side of the basement pulling strings. In all, only four lights worked to illuminate the basement.

The floor was dirt, and huge brick pillars about six feet tall and two feet wide were spread out everywhere. Three of the outside walls were brick, one was smooth and painted white. The boys walked cautiously between the cobwebs and used the flashlights to see in dark corners or the dirty basement. In the back corner of the basement was the other set of stairs. They were barely visible because the door to the outside did not have a window by it. On one side of the large basement there were rows of shelves with old jars of pickles, jelly and other strange canned items. Cobwebs covered most of the jars. Near the other set of steps there were old tires and rims, and what appeared to be old auto parts. In the middle there was a huge pile of junk.

In all, it was a pretty grim place. Pete was disappointed in the whole basement and stood near the middle and made a circle with his flashlight. Davy was near the edge where there was a stack of old furniture. He saw a bedpost, dresser drawers, old end tables, lamps, and odds and ends. "It doesn't look like anything good is down here," Pete said with a sigh.

Davy, who was suddenly pulling at something said excitedly, "Wow, look at this." Just then, something seemed to roll out on the floor with a thud, and Davy jumped back. His flashlight beam was on it and he exclaimed, "What is it?"

Pete made his way carefully over to where Davy was standing and his beam went over to the large silvery round object lying on the ground. "I think it's a spittoon." Pete bent down and poked it with his finger. It rolled about three inches on the uneven dirt floor, and then rolled back. As it did, a large clanking sound came from inside. Pete and Davy both moved to where they could see the opening. Inside appeared to be a heavy metal object. Pete put his beam directly into the opening to get a better look. Without hesitation, Davy bent down and slowly put his hand in the opening. Pete was stunned that Davy was so brave. He wasn't going to put his hand in anything so creepy.

Something touched Davy's hand and he pulled it out quickly, shaking it as if there was something attached. "What is it?" asked Pete in an excited voice. "Nothing," replied Davy and he got down on his belly. He flashed his beam of light so he could see what was inside. Again Davy put his hand cautiously into the large opening and pulled out a dusty metal object. It looked like a sun with jagged edges pointing out in all directions, and there was a bubble in the middle. The bubble had a free floating needle that looked like a compass. Davy turned the object over in his hands and wiped off the cobwebs. The bottom of the object was flat and smooth with markings on it, and the other sides looked like rays of the sun. The large ray at the top had the letter N, and the bottom ray that was fat and covered the entire bottom had the letter S. W and E were on the sides. It was a compass, but not like any compass either boy had ever seen.

Davy looked up at Pete and said, "This could be a treasure, what do you think?" Pete slowly shook his head and said, "Yeah, it could be, I wonder what else is in here." With that, both boys aimed their flashlight beams toward the large pile of junk that cluttered the furniture, shelves and floor.

Davy moved his beam in slow deliberate movements as to see everything. He moved larger objects so he could see what was behind everything. He got on the ground to look under large pieces of furniture and climbed over other objects to look behind them. Pete kept his beam moving quickly, he was becoming nervous.

Finally, Pete stopped and stepped back to observe Davy as he went about looking for something more. Pete was getting agitated with all the darkness and shadows and was really quite ready to move out of the basement all together when again Davy made another discovery.

"Hey, look at that, I think that may be something cool," and Davy reached over a bed post on his tip toes to reach a strange looking object on top of a shelf near the back of the stacks. His hand came to rest on a cold piece of dusty metal. He slowly and carefully pulled it toward him until he could wrap his fingers around the edge of the metal. It was heavy and he had to grab it hard to pull it off the shelf.

When he got it down, he turned it over in his hands and wiped the cobwebs off. It was heavy and about the size of a pie plate. It looked like a wedge, one side was round while the other angular. It appeared to have an eye piece, but when you looked through it, nothing appeared to be bigger. It did have a dark blue and a dark red piece of glass no bigger than a finger nail hinged to the middle. If you looked through the eye piece, you could move the pieces of glass in front so you could see one or both as you rotated them. "I think this is a treasure. Do you know what it does?" asked Davy.

Pete moved over to where Davy stood and reached for the object. Davy handed it off and Pete noticed right away how heavy it was. It must have been made a long time ago. He rubbed his hands along the rounded edge and noticed it had calibrations, and numbers. "Wow, this really is a treasure," said Pete. Davy gave a happy sigh and said, "You want to ask your Pops what these things are? Do you think he knows what they are?" Pete looked at Davy's face and wondered what kind of story he would write about these new fascinating objects.

A rattling noise came from the pile of junk and both boys jumped. "What was that?" Pete said in a low voice. "I'm not sure," said Davy as he flashed his beam at the pile of junk. It seemed that both boys were staring at different locations in the pile when from the very top of the heap a large round object began to wobble. Both boys looked up at it the very same time. Neither boy moved or spoke. Pete was frozen with fear.

The wobbling object was a large dirty, bowl. It turned, and then seemed to teeter towards the edge of the shelf. Then in what seemed like slow motion, the bowl fell to the dirt floor and landed with a loud crash. It bounced, and then shattered into fragments. Both boys screamed and jumped back, Pete falling on his butt on the hard ground. "I'm done here," he yelled and grabbed for his flash light. He rolled onto his knees and scrambled for the steps.

Davy looked around, shaken by the falling bowl. His flashlight was in his hand and so he flashed the beam toward the top of the shelf. He thought he saw a small silver-hared animal with a long rope-like tail. He walked closer and saw it was a small rat and his eyes flashed bright green in the beam. Davy looked toward the steps to tell Pete it was just a rat, but Pete was gone. Davy shook his little head and wondered how a guy so cool like Pete could be such a ninny. I mean really, look at me he thought, and it brought a smile to his face to think he could be tougher than a guy like Pete.

He gathered up the two mysterious treasures and turned off the lights. He ducked carefully under the cobwebs and started toward the steps. As he walked up the steps, he saw Pete standing just inside the door. "What a cool basement, can we come back and look around again?" Pete just sighed, and nodded in agreement.

Pops rolled the mysterious object in his old, weathered hands. "Boys what you got here is a sexton, it gives you the altitude of the sun in the sky. It was used by sailors to give location at sea. It helped them navigate before we had high technology, satellites and computers." Both of the boys, wide-eyed stared at Pops. The elderly gentleman with silver hair and a thin face, eyed the boys as they eagerly wanted more information.

"What about this?" Davy said as he held up the sun-like object. "Do you think this is a tool for location too?" Pops sat up straight and put the sexton down before taking the strange looking object.

"Hmmm," he said as he slowly rolled this object between his hands. He shook his head and then rolled it again. "All I can really say about this is that it has a compass in the middle, the rest is a mystery to me."

The boys looked at each other and then Pete came very close to his grandfather and put his arm around his shoulder, pretty much the way a small child does before he ask for something. "Pops, could I keep these treasures, I mean if you're not using them I'd like to keep them in my room for awhile?" Pops looked down at his hands and twiddled his thumbs. He shook his head and then looked at Pete square in the eye, "They are very important and serious objects, not to be taken lightly, and if you leave them out in the rain or misplace them, what am I going to tell your grandmother?" Davy jumped forward and put a hand around the other side of Pops, "Sir, I promise I won't let him do that," he said while he shook his curly head with such force that you would think the curls would come tumbling off. "If he so much as puts them down and walks off, I will personally beat him up, these things are so cool."

With a smile across his weathered face, he looked first at Pete and then at Davy and said, "If you boys promise to take care of this stuff, then I guess I could let you keep it in your room, for a little while." Pete let out a scream of joy and Davy scooped the items up quickly and put them in his shirt, as to hide them from Pete's grandma and mom. The boys made a bee-line upstairs and a door slammed to the bedroom.

Pops chuckled to himself as he walked into the kitchen. Pete's mom was feeding his little sisters apples and crackers at the table and they looked up as Pops walked through the door. "What's so funny?" asked Pete's mom. "Oh, those two boys, they remind me of myself when I was that age." He shook his head and filled a glass with tap water and drank it all down at once. "His father was never like that, always so serious. Seems that boy lost his sense of adventure and never found it again, but that Pete, he is something," and his voice trailed off as he gazed out the kitchen window.

Upstairs, Pete held up the object with the compass in the middle and stared closely at it. He brought it up all the way to his nose. Davy watched Pete and then said, "Do you think it can find the North Pole?" Pete looked at Davy with a question on his face, "Why the North Pole?" "Because of the compass, it has a compass in the center."

Pete held the object out and looked at it, "Yeah I guess it could, but I think there is more to this than just a compass."

Davy took the other object and ran his finger over the smooth, rounded edge. "I bet some ancient warrior used this to find his way. I saw one of these in a book on ocean voyages, it has something to do with locations and ships, and ..." Davy rambled on and Pete was lost in his own thoughts.

The two boys exchanged objects and discussed the importance of their finds, although neither boy had any real clue as to what they were or what they were used for.

CHAPTER FOUR

THE FIRST DAY

School started the middle of the third week of August. New shoes, book bag, and a lunch box were just not enough to make the day special anymore. Pete walked to the bus stop with his little sisters and his mom. This just happened to be the day his little twin sisters, Madelyn and Ashlyn started Kindergarten. Apparently new shoes, a book bag and a matching lunch box does it for someone who is five.

"Stand still and smile," said Pete's mom as she snapped another picture. Pete, the good kid, smiled again for another picture. Seventh grade seemed to be just another grade in the growing up process, and Pete wasn't happy to end his summer and go back to school.

"I promise I'll watch them on the bus," said Pete "they will be just fine." Pete put his arm around Ashlyn while the other little girl hugged her mom and cried. The bus came and with much effort from Pete and his mother, the little girls were safely on the bus.

Pete sat in a seat near the front and put one sister on his lap while the other sat in the seat next to him. He wasn't thrilled to be the care taker of these little girls, but he did it because he was a good brother.

The bus was almost full when Pete saw the most amazing thing. At the last stop, a girl with bright green eyes and curly auburn hair walked up the steps and looked around. A weird sensation of super-slow-motion struck him as she looked around and gave a shy smile. She focused her gaze right at him and outright grinned, and then took the seat right in front of him.

Pete's whole world went silent as she sat down, turned her head and said, "Hi, I'm new." Pete heard the words flow swiftly out of her mouth, but could not reply. Ashlyn, the twin sitting on Pete's lap started to giggle. Madelyn, the other twin reached up and poked Pete under the chin. "I think his brain is stuck," and she poked him again. Pete, hearing his sister's words and feeling the sharp pain of a little finger poking in the soft, tender skin under his chin, gave a jump, and shook his shoulders. "Hi, I'm Pete and I'm not new," and he gave a weak smile.

What a goofy thing to say. How could he have turned into such an ogling fool in such a short time. He was being so good watching out for his little sisters and now he was an ogling fool.

Madelyn took over and started talking for her older brother. "I'm Madelyn and this is my sister," she smiled and held her sister's hand in the air, as to introduce her. "I'm Ashlyn," said the other and smiled, sticking her chin into her chest.

"Aren't you the cutest little twins," said the girl in the front seat taking the hands of one of the girls. "My name is Amanda, and this is my first day at this school."

The little girls giggled and smiled. "This is our first day too, maybe you will be in our class," and the giggling continued.

Pete, who suddenly realized his little sisters misinterpreted the situation jumped into the conversation, "I think that they think because you are new you will be with them in their class because they are new." Pete sat back and thought about what he was saying, "I mean, they seem to think that if you are new, you must somehow be in kindergarten." He put on a weak smile followed by a huge sigh. When he realized everyone was looking at him, he brushed his hand over his eyes and looked out the window.

"Pete is our brother, and he is so smart," smiled Madelyn sitting next to him, holding his hand.

Ashlyn looked at the girl sitting in front of them and said, "I think you're pretty" and gave a little giggle. "I think you're prettier than my mommy," and she giggled again The two little girls passed a look between them, and fell together giggling.

Pete seemed to hear all of this and his face turned bright red. He could feel his face get hot, even his ears were burning.

Amanda's face turned pink, and she gave a nervous giggle. She turned immediately to face the front of the bus and avoid the gaze of the little girls.

When the bus stopped, the intriguing girl jumped up and was the first one off. Pete maneuvered one girl off his lap and stood taking both of their hands. Carefully he got them off the bus and walked them to the location of the kindergarten teacher, who was busy greeting all of the young children.

After he finished dropping them off, he looked around to see if he could locate the girl who was sitting in front of him. She was nowhere to be found. He shrugged his shoulders and walked toward the seventh grade classrooms. From the corner of his eye he saw Billy giving him an evil look. He hadn't seen Billy since the day he pelted him with water balloons. Truthfully, he had been avoiding him, but seeing Billy glare at him made him smile. He put his head down and walked inside, not wanting to give Billy a reason to jump on him right there.

He knew the name of his teacher and the room number, so all he had to do was walk in and look for his name on one of the desks. He already knew that Davy was in a different classroom, but he would be able to see him at lunch. He said hi to a few of his friends as he made his way around the room looking for his name. When he found it, he put his backpack on the desk and slid gracefully into the chair. As soon as he was sitting still, he started to stare off thinking about the new girl and wondering if he would see her again. He frowned when he thought about what a dork he must have looked like.

When the teacher finally walked in, and asked the kids to put their backpacks on the hooks in the back of the room, Pete stood up and turned around bumping into the person sitting right behind him. "I'm sorry," was all he could say when he looked into the face of the same girl.

"That's okay," and she smiled at him.

"Come on Pete, move it," said the boy behind him. Pete began to move his feet, he hung up his pack and sat down without realizing that he even moved. When he was sitting in his desk, he became frozen thinking she was right behind him. What to do now? Should he just sit there and pretend he didn't see her? No, that was stupid. Of course he needed to turn around, smile and act normal. But what if he couldn't talk or worse yet, what if he acted like the big geek he did on the bus. Stricken with fear, he just froze.

It was Ms. Nagel who came to his aide. "Ladies and gentlemen, I know that most of you already know each other, but we have a couple of new students at our school. Jake Brown, please stand up. This is Jake and he is coming to us from a school in North Carolina." A tall, lean boy with red hair and freckles stood up, bowed to the class in a comical way, and then sat down very quickly.

"And this is Amanda Johnson, Amanda please stand," and the girl behind him stood up. This gave Pete the excuse he needed to turn around and smile at her. "Amanda comes to us from across the river in Ohio. Make sure you introduce yourselves to our new students and make them feel at home."

Pete extended his hand to Amanda, who with her amazing smile, took his hand and shook it. "I'm Pete, I sat behind you on the bus," said Pete. "Yeah, I know, it's nice to meet you again."

From the corner of Pete's eye he caught a sharp movement of another hand abruptly approaching the front of Amanda. "Charlie McMullin, at your service." Amanda immediately shifted her gaze to Charlie and dropped Pete's hand to take his. She gave a smile and said, "Nice to meet you Charlie." Charlie shot Pete an evil glance and looked back to Amanda. "I'm sure we're going to be good friends. I have lots of friends I want you to meet, and if you want to, you can sit with my friends and me at lunch," Charlie said. "Thank you, I don't really know many people, and that was nice of you to ask, maybe I will," and she continued to smile at Charlie.

Someone behind Amanda tapped her on the shoulder and she turned quickly around to see who it was. When she looked away, Charlie glared again at Pete, "You're dead meat," he said in a low growl.

"Oh yeah, first you have to catch me, and you can't catch me if you're running in the other direction," said Pete as a huge smile erupted across his face. Charlie's glare grew more intense, but before he could say anything else, Ms. Nagel called the class back to attention.

The boys gave each other another long, mean glance, and then turned to the front, but not before Amanda could see the tension between the two boys.

At lunch, Pete waited at the opening to the Cafeteria for Davy. He happened to notice Amanda sitting with a group of girls, some of whom he thought to be friends of his. He was relieved to see she wasn't sitting with Charlie and his buddy Billy Martin.

When Davy's class finally arrived, Pete had already begun to eat his sandwich. Davy grabbed a tray and together they sat with a few other boys. "Did you see that new girl?" one of the boys said. Pete stopped eating and started listening. "Yeah, she's really pretty, you know, as girls go," said another.

Davy noticed that Pete stopped eating, and so he stopped and started listening too. After a few minutes, Davy shook his head in disgust and said, "They're talking about a girl. What's so great about a stupid girl?"

Pete looked at him with a strange expression on his face, "Have you seen this girl, have you talked to her?" asked Pete

"Pete, you need to snap out of it, this is seventh grade. No one likes girls, I mean really likes girls until at least high school, and then still they have cooties," said Davy. He took the last bite of his sandwich and tore open his chips. Davy was one of those people who could only eat one type of food at a time. First he ate his sandwich, then his chips, every single one before he went for the cookies. After he ate everything in his lunch, he would look around the table to see if anyone had anything left he could eat. Davy could eat and eat and still stay skinny.

Davy looked around nervously, he saw Billy Martin glaring at him. He looked over at Pete and said, "That stink-pot Billy Martin is staring at us and I think we should.." and he suddenly stopped talking when he noticed Pete had stopped eating and put down his sandwich. He seemed to be watching a group of girls. Suddenly he stood up and slammed his hands on the table. Davy looked over in the direction of his gaze and saw Charlie standing next to the girl with shiny, curly hair. He moved a stool over and sat down next to her, pushing one of the girls out of the way. Pete just stared, and slowly sat back down.

"Not fair, I saw her first," he said. Davy just shook his head and continued to eat.

After lunch, break immediately followed and so when you were done eating, you threw your trash away and went outside. If it was raining, you went to the gym. It just happened to be raining, and so when everyone was done, they were all gathered in the gym. Since it was the first day of school, no clicks had really formed, and so the students were all just standing around the perimeter of the gym floor.

Pete and Davy got to the gym early, and Pete was waiting for Amanda to come in so he could talk to her some more. Even though talking to Amanda was rather difficult for him, he wasn't going to let Charlie get to her first. "What do you want to do after school? My mom is taking my little brother to another specialist, and so I have to go to Mrs. Elders and wait for her, but Mrs. Elders probably wouldn't mind if you came by. She really likes you, and she would probably pay us to do some chores for her." Davy was rambling on. Pete nodded his head in agreement, when she entered the gym. He pushed himself away from the wall, and stood looking in her direction. She wasn't with Charlie anymore, and so he had to move fast before he came back. Without a word to Davy, he started walking quickly in her direction. Stunned at the effect this girl had over his friend, Davy started after him. He was almost next to him when Pete suddenly stopped. He noticed that Amanda was being surrounded by girls from nearly all sides. He turned around abruptly so she wouldn't notice his sudden approach. Davy, caught off guard did a 360 degree turn. How was Pete ever going to talk to her if she always had someone around? What was he going to talk about anyway?

Pete stood in the middle of the gym floor scratching his chin and shaking his head. Davy stood next to him, and quite out of the blue started scratching his chin and shaking his head too. Together they looked like a pair of bobble-heads.

"Pete," said Davy, "what are we doing in the middle of the gym? Didn't you want to talk to her? Why did you just stop?"

"Shhh, she could hear you," said Pete. "Is she looking at me, I mean us?"

Davy turned and looked directly at the girl Pete had claimed was so beautiful. She did kind of have that pretty quality some girls have, but she wasn't by any means beautiful. Davy stared at her until she suddenly looked up and glanced back. Davy's body seemed to jerk back as she looked back, and he turned around rather quickly and said, "yep, she's looking right at us." "Oh great, what do we do now?" moaned Pete.

Saved by the bell, the entire class headed off to the doors and down the hall back to the classroom. Pete was falling into deep thought. How was he ever going to get to talk to Amanda, and not look like a fool, or a geek doing it?

The rest of the day seemed to go way too fast, and before Pete had a chance to make a plan, the bell rang to leave school. Pete drug his backpack over his shoulder and was off to find his bus, and keep an eye on his little sisters.

Pete stood outside his bus wandering when his sisters would arrive, when the bus driver motioned for him to come in. "The kindergarten gets on the bus first, your little sisters are right here." Pete looked in, and there they were, carbon copies of each other.

"Thanks," said Pete as he patted them both on the head, one at a time. "I'm going to sit behind you since there aren't that many people on the bus," and he plopped into the seat behind them.

When he looked up, there she was, sitting in the seat across the isle. There was no one else in her seat, and she looked over at Pete. Pete swallowed hard and pushed himself to the edge of the seat. It's now or never he thought, and launched a conversation.

"How did you like school today?" he said. She smiled and scooted to the edge of her seat to respond. The chat lasted all the way until Amanda got off the bus, and she waved good bye to all of them when she left.

Pete watched her as long as he could, and when she fell out of his sight, he pushed himself back in his seat and gave a happy sigh. If this is what it was like to have a girl crawl under your skin and consume your every thought, keep you from eating, and possibly sleeping, then he was doomed. Shoot me now he thought, because tomorrow was going to be the same as today.

He closed his eyes, and was falling into despair about the coming day and not spending time with Amanda, when a little head popped over the seat. "Are you going to sit with Amanda tomorrow on the bus?" a little voice said. Then another head popped up, and another little voice said, "I think she would like to sit next to you," and the giggling began. And then it hit him, like a ton of bowling balls. Everyday Pete would have his time with Amanda on the bus, and he could sit next to her and talk to

her everyday before and after school. A big grin cornered his lips and the bus stopped. "Come on, it's our turn to get off," said Pete, and he helped the little girls off the bus.

CHAPTER FIVE

WHAT GOES AROUND COMES AROUND

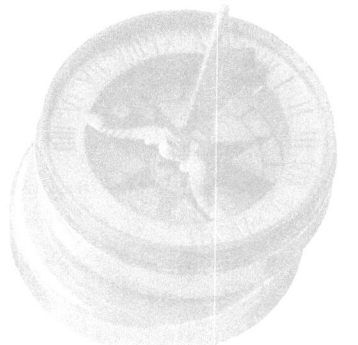

It rained most of the day, but it appeared as if the rain had finally ended. After the first day of school, Pete was on his way to Mrs. Elders' house to see Davy and hopefully explain the 'Amanda thing'. No girl was ever going to come between their friendship, except maybe Amanda. No! No! Never would any girl come between them, and that was why Pete was walking so fast and not paying any attention to the boys he walked right past. He didn't see Billy or Charlie at all when he passed them on the corner by the Quick Shop. He didn't hear when Billy called out to him, or when the boys suddenly disappeared behind the bushes the moment he stepped into Mrs. Elders' yard.

They watched as Pete knocked on the door and then went inside. The plan was concocted right on that very spot, behind the bushed in the yard across the street. Pete and Davy had been hiding and avoiding Billy and Charlie, and this was the first chance to get revenge. Today, revenge was served with ketchup, mustard, and chocolate sauce; and it was going to be messy.

Pete went inside the house and the first thing that hit him was the smell. It wasn't of flowers or air freshener like his house, which was much older, but the smell was like old coffee that stayed on the burner too long. As he moved into the kitchen, the smell grew more intense. It wasn't like the house was messy, but the smell was becoming overpowering.

"Hey, could we go out back and hang out?" Pete asked. Mrs. Elders' agreed and followed the boys to the backyard.

"I was wondering if you boys wouldn't like to make some money and rake up those dog-gone cottonwood leaves. They fall all the time, and if I don't get them up, they will kill the grass," she said to them.

Davy put his hands on his hips and was about to make a bargain when Pete suddenly said, "sure, we can do it." Davy gave Pete a quick glance that Mrs. Elders' totally missed, and she shuffled off to the shed to get the rakes and a trash can.

"Hey, I was going to ask her how much she was going to pay us. Why did you have to go and say yes so fast," Davy barked at Pete.

"Listen, she'll pay us, she always does. I just don't feel like asking for money," replied Pete.

Davy gave Pete a long, hard look and then plodded off in the same direction as Mrs. Elders'. She handed both boys a rake then smiled at them, and went inside.

Pete went to the side of the yard and pushed his rake outward to pull in as many leaves as he could. His plan was to make small piles and then bring the trash can to the piles. He started to talk as he began to rake. "You know that new girl, the one at lunch?" he said to Davy.

He heard Davy make an "Uh hu" sound. "Well, she rides my bus and I had a chance to talk to her. She's really a nice girl," he said. Pete was trying to choose his words carefully so he didn't offend his friend by his behavior today, or the fact that he was just about to breech the boy-code of 'NO GIRLS' as friends.

"Did you get a chance to talk to her at all?" asked Pete. He stopped raking and looked directly at Davy.

Davy seemed to be making a tractor noise as he imagined himself on a huge tractor raking in the leaves. Pete looked at him, and Davy said, "No, I just saw her in the gym, why?" and he went back to his motion with the rake and the noise.

This is going to be harder than he thought. Davy no more wanted to talk about girls than he wanted to hang out with Billy Martin. He had to have a better approach, and so he decided to take it in a new direction.

"Are there any new kids riding your bus?" he said.

This Davy seemed to hear quite well. "Nope," he said and made a nice tidy pile of leaves. He moved to a spot by the big peach tree and started making another small pile of leaves.

"Davy," Pete said and stopped raking, "I want you to meet her, Amanda."

Davy didn't even look up he just moved his rake in the direction to catch the most leaves. It seemed as if he already knew that this girl was going to change everything. He didn't even look up when the balloons started to sail across the yard.

The first one hit the peach tree, and by the looks of the contents, it was full of mustard. A yellow splat covered the whole side of the tree trunk. The second one hit Pete square in the back of the head and bounced to his shoe where it exploded with dark brown chocolate sauce everywhere. Pete yelled out, "Ow, what the heck?" and he realized that he was under attack by the 'gruesome two-some.'

Davy looked up to see what Pete was yelling about when a balloon filled with ketchup hit him square in the middle of his shirt. "Hey!" he yelled and took his rake and put it in front of him to protect him from other balloons.

The battle had begun, but this was far worse than either boy could have expected. Balloons filled with bright yellow, red and brown were breaking everywhere. Pete got hit smack center in the head and was dripping ketchup off of his nose. Davy was swinging at the balloons, and as they exploded on his rake, the debris would splatter on his cloths and the tree and grass.

"Where are they? Can you see them?" yelled Pete as he maneuvered around the tree next to Davy.

"Yeah, I see them, they're right by the edge of the house," yelled Davy as he tried to get close to Pete behind the trunk of the tree.

Balloons were hitting everywhere around them. Their faces, shirts, legs and bodies where red, yellow and brown from the massive amounts of balloons.

"Hey, I got an idea, but it might be a little dangerous, you in?" said Pete hurriedly.

"It's got to be better than this."

"Okay, I say we chase them down with our rakes and steal their balloons and nail them. You sure you're in? They may get mean." said Pete.

Davy ducked as a balloon sailed over his head and hit the fence behind him and splattered all over it. "Oh yeah, I'm in."

"Okay then, we make a lot of noise, we scream as loud as we can as we run their way, that may distract them and cause them to run off. Can you do that?" said Pete as a large balloon hit the tree over his head splattering ketchup all over the branches.

"On the count of three, ready? One, Two, Three," and they were out from behind the tree and running to the edge of the house screaming as loud as their lungs could bear. They covered the yard quickly and were near the house when the other two boys realized they were coming for them. Startled, they jump back and started to run leaving their stash of balloons in a bucket on the side walk.

When Pete and Davy saw them leave, they made a pocket in the front of their shirts and filled them with the remaining balloons. Then Pete went around the side and Davy went around the back.

Billy stopped running when he got to the front of the house and realized that he had left the stash of balloons near the back of the house. He grabbed for Charlie's arm, and yelled, "Wait, we have more balloons. Let's get 'em," and together they turned just in time to have an onslaught of balloons hit them. They were hit in the head, abdomen and shoulders. There were spats of red, yellow and brown all over their clean shirts and shorts.

At first, they were surprised by the attack, and then Billy yelled, "Get them!" and they ran towards the boys holding the balloons.

From the front window, Mrs. Elders couldn't believe her eyes. Red, yellow and brown balloons were being burst all over her yard. "This has got to stop, NOW!" she yelled as she stormed out the back door and grabbed the spray nozzle and spun it on the end of the hose. She turned on the water full blast, and pulled the hose around to the front of the house. When she arrived, Billy had Davy in a head lock who was kicking at his feet, and Pete and Charlie where hitting each other with open hands, like girls fighting.

She turned on the hose and started to spray Billy and Davy first. She blasted Billy, the big bully, whom she really didn't like, right in the face. As she sprayed she got closer and the power of the sprayer gained momentum. By the time she was right in his face with the hose, aiming it towards his big nose, he let go of the littler boy and was trying to cover his face with his hands.

"Step back you big bully!" she yelled over the sound of the hose spray, and continued to spray him in the face until he relented.

When he did, she turned the hose on the other two boys. "Stop!" she yelled, and immediately the two boys stopped and looked at her.

All four boys were wet and covered with yellow, red and brown in their hair, cloths and body.

She let go of the nozzle and looked at the four boys. She took a deep breath and in a calm, low voice of someone in authority, she said, "Clean this mess up NOW!"

The boys looked at each other, and without a word they started picking up the broken balloons.

"What are you going to do with the other mess?" and she pointed to the spots of red, yellow and brown on the sidewalk, the bushes, and her house.

"We'll clean it up," said Pete and he walked slowly towards Mrs. Elders and took the hose out of her hands. "You guys pick up the balloons in the back of the house, and I'll spray the house and sidewalk."

"What about the leaves I asked you to pick up?" she said looking at Pete and then at the other boys.

"We'll do that for you, don't worry," said Charlie who had a very sheepish look on his face. "I'm sorry we caused such a mess," and he motioned to the other boys to follow him to the back of the house and start cleaning up the balloons and the leaves.

Davy and Billy, who were giving each other the evil eye moved to the back of the house behind Charlie. "You little .." but before Billy could finish his insult Charlie shushed him and said, "Billy, we don't need to have her call our parents, just be quiet and clean up."

When the boys reached the back of the house, Billy stopped and gave Davy a mean look. "I'm not picking this up, I'm not picking anything up," and he put his hands on his hips. Charlie was embarrassed to have been caught, and he didn't want to get in trouble at home. "Billy, you need to help us."

Billy was glaring at Davy, "You little monster, this is all your fault."

Davy stopped and looked into Billy's red, swollen face. He was ready to respond when Mrs. Elders walked back to where they were. "I know all of your names, and I'm really thinking I need to call your parents." Davy looked at her and bent down to pick up the broken balloons. Billy bent down and picked up a balloon, and grumbled under his breath. He barely made an effort to clean up, and spent a lot of time kicking and punching thin air.

After 45 minutes of cleaning, the yard looked great, and every inch of color was washed away. Billy glared at the other boys and walked up the street. Charlie stopped and knocked on the back door.

"Mrs. Elders, I'm sorry if I caused you any grief, it was not my intention to include you in our escapade, I apologize," said Charlie.

Mrs. Elders smiled at Charlie and said, "What a nice young boy you are, come by and I will let you mow my lawn. I'll pay a good price for your services," and she smiled and patted him on the head. He turned and walked toward Billy.

Pete and Davy shot him a nasty look before he turned, and he smiled wide for both of them to see. "What a jerk," said Davy under his breath. Pete looked at Davy and nodded in agreement. They were the ones who were attacked and Charlie McMullin was being labeled as a 'good kid' who helped clean up. Heck, he was the one who helped make the mess in the first place.

Charlie met up with Billy and said, "Hey, nice of you to take off without me." Billy shrugged his shoulders and said, "What was that all about, you kissing up to that old broad?"

Charlie stopped walking and pulled on Billy's arm to stop him from walking. "Are you kidding me, we messed up that old woman's yard, and she didn't really deserve that." He glared at Billy who looked at his shoes.

"Who cares? She's just some old lady anyway," said Billy.

"What kind of attitude is that? Old people aren't just nobodies that you can mess up their yard and not care. I like old people. I think they are interesting, kind, and I would like to help her if she needed it," Charlie said. "Don't you have a grandma, grandpa that you like?"

"I don't really care for my grandma much, all she does is complain, and I think she's mean," said Billy. "My grandma lives in one of those places, and I hate to go there. My mom makes me go and I hate it. I hate old people," Billy said, and he started walking up the street with his head down.

Charlie let out a huge sigh and started after him. He didn't think Mrs. Elders was beyond asking them to help clean up, since it was a huge mess that they had created. He felt sorry that Billy didn't care for older people, because he really did like his grandparents.

Davy and Pete were a mess. Their cloths were soiled, their hair was clumped with mustard, ketchup and chocolate sauce, and their tennis shoes were soaked. They stood outside Mrs. Elders' back door with their heads down. When she opened the door Davy looked up and said in his sincerest voice possible, "I'm really sorry about this, but we didn't start it." His blue eyes begged forgiveness.

She reached down and put her hand under his chin lifting it up to meet her gaze, "I saw the whole thing, I know you didn't start it." She smiled at him and pulled out some dollar bills from the pocket in her apron. She put $2.00 in his hand and said, "This is for picking up the leaves," and then put $2.00 more in his hand and said, "And this is for standing up to those two bullies." Davy dropped his gaze and a barely audible "thank you" came out as he pressed against her. "You're welcome," she said and patted the gooey mess of curls.

Pete stood and watched the two of them. He suddenly realized that this woman knew Davy needed her attention, and she was more than willing to give it to him. He wondered how deep their friendship really went.

After a long moment, she looked at Pete and pulled some more money out of her apron. She handed Pete a wad of bills and nodded at him, and he nodded slightly back.

"I probably should go clean up," Pete said, and turned to walk out of the yard.

Davy pulled away from Mrs. Elders a little embarrassed, and walked Pete to the end of the yard. "Hey, you want to hang out after school tomorrow?" asked Davy.

"Sure," said Pete, and he walked towards his house.

CHAPTER SIX

ALL MESSED UP

Pete was sitting on the bus, his little sisters sitting in the seat in front of him, and he was patiently waiting for Amanda to get on. He was taking note of where her stop was so that he could ride his bike to her house.

It was next to the last stop before they got to school. She boarded the bus with a big smile and sat in his seat. "Hi," she said shyly. She looked directly at Pete, smiled and then said, "So what's up between you and Charlie?"

Pete was taken off guard, and he immediately turned and looked out the window. Amanda had struck a nerve, and he didn't know how to respond.

Scratching his head, he returned her gaze and said, "We're not friends."

"I noticed," she said. "Why aren't you friends?" She looked at him as if he would just spill information.

After a few minutes of very uncomfortable silence she said, "I just don't know why the two most popular boys in the seventh grade aren't friends?" Her smile and her gaze remained steady and Pete knew he would have to give a response. "Who says I'm popular?" replied Pete with an embarrassed look. "I do," said Amanda. "Everyone talks about you, and how smart and, and,...just popular," she stopped before mentioning handsome or athletic. She could see that he was uncomfortable with the conversation.

"Well, I'm not really popular, and I don't like Charlie McMullin. He's friends with Billy Martin, and he's just mean."

Amanda nodded in agreement. She had heard Billy was mean.

When they exited the bus, and his sisters were safely on their way to their kindergarten class, Pete caught up with Amanda and walked next to her into the classroom. He didn't want Amanda to get the wrong idea about him not liking Charlie. "Hey, I don't want you think that I go around making enemies with people, its just that Charlie never liked me either." Amanda kept walking as she said, "Seems like a stupid reasons not to like someone, just because he never liked you, or he has a mean friend." When they entered the classroom, the desks were all different. Instead of rows, the desks were neatly placed in groups of four. The students were standing on the outsides looking for their names. Several students had found their new spots and were sitting quietly, while others were still milling around.

Pete's stomach sank. What if he wasn't sitting next to Amanda anymore. What if he had to sit next to Charlie, or Billy? He groaned and hung up his back pack and then started to look for his name. He heard Amanda's voice calling his name from across the room and looked in her direction.

"Pete, Pete, you're sitting across from me," she said. Relief overcame him as he moved towards her. When he reached the desk, he was in shock to see that his seat was next to Charlie's, who was already sitting in his seat, smiling and making small talk to Amanda.

"Great, just great," he said as he slid into his chair. Charlie and Amanda stopped talking and looked at him. Pete tried to give a pleasant smile, but he was overcome with jealousy.

The fourth person at the table was Elizabeth Stanley. She was a small, mousy girl who barely said anything. Pete wouldn't mind having three mousy girls, just not Charlie McMullin.

When the teacher started the class, Charlie turned and gave Pete an evil glance, and Pete returned it immediately. Amanda didn't miss the animosity between the two boys.

Charlie leaned in and whispered to Pete, "I know you hate sitting next to me, maybe you should ask the teacher if you can move," and mean smile crossed his lips.

"You wish," and he insincerely smiled back. Pete was not going to move a single foot away from Amanda no matter how close Charlie was sitting next to him.

The morning was long and tedious as the class reviewed math functions. At midmorning, the class got out their science textbooks and began a project. Each group was assigned a scientist to research and then they had to, in turn, give a presentation to the class on that scientist. Pete's group was given Johannes Kepler, a mathematician who studied the planets.

Charlie seemed happy to get this person, but no one else in the group had ever even heard of him. When the teacher stopped explaining and let the groups talk, it was Charlie who spoke first.

"This is great," he said looking directly into Amanda's eyes. "I know all about Kepler. He came up with the laws that govern planetary motion. His first law says...." and that was all he got out before Pete interrupted, "So how do you know so much about Kepler?"

Charlie turned his body and leaned on his elbow to stare directly into Pete's eyes. "I went to Science Camp three years in a row. My mother insists that I read books on scientists, along with bibliographies on historical figures. So while you're out throwing water balloons at other kids, I'm studying and improving myself." He said it so smugly, that Pete couldn't form a comeback.

Charlie moved deliberately to face Amanda and continued, "I know a great activity for his first law of planetary motion which says that all planets have an elliptical path. You take a piece of cardboard, two straight pins and a piece of string tied in a loop. You put the pins about 1cm from each other on the cardboard and put the loop around them. You take your pencil, pull the string taunt and draw straight. It makes a flattened circle called an ellipse. Kepler says that all planets have an elliptical path, and that the sun is at one of the pins." He stopped briefly to see if everyone was listening, and when he did, Pete jumped in again.

"Yeah, so big deal, what does it really mean?" He said curtly to Charlie.

"Yeah, it is a big deal," he said back. "It means that we are closer to the sun at one point in our orbit and farther at another," Charlie replied as he looked around the small group. Everyone seemed to be amazed at his knowledge on Kepler. He liked looking smarter than Pete, who was also very smart.

It was Elizabeth who interrupted, she raised her hand, as if she wanted to ask a question, and when everyone looked at her, she asked shyly, "Does that mean when we are closer to the sun we are having summer, and when we are farther away we are having winter." She put her hand down and looped it through the other in a nervous gesture.

The group looked at her for a moment, and then Pete jumped in, "Actually, I don't think that is why we have winter and summer. I think it has to do with the earth being tilted."

Charlie waited for his moment and then added, "One would think that if you were closer to the sun, then of course you would be warmer, but Pete is right, that is not why we have a change in winter and summer. January is when we are closer to the sun, and it is in the middle of winter. July is when we are farthest away and this is when we are having summer. Distance from the sun has very little effect on our seasons."

Amanda looked at Charlie in amazement, and Charlie's face went pink. Pete wanted so badly to trip him up, but Amanda finally spoke up. "We probably need to do an internet search and read about this guy Kepler, and then we can decide who's going to do what part of the presentation."

They all nodded in agreement, and then Pete said, "We could look him up right now in our science books and see what they say about him in there." At that, all four took out their books and thumbed through looking for Johannes Kepler.

"I think I found him," said Elizabeth, "I'm on page 295." With that, the other three turned quickly to that page. Before Pete could utter a work, Charlie started reading out loud, "Johannes Kepler had a strong mathematical mind, and working on the findings of Tycho Brahe, he came up with the three laws that govern planetary motion." Charlie took a deep breath, and was about to continue reading when Pete said, "we can all read." Ignoring Pete, Charlie began again, "The first law of planetary motion states that all planets follow an elliptical orbit around the sun, and the sun is at one of the foci, which are reference points for the ellipse. The second law says that each planet covers equal areas in equal time. The third law states the planet's orbital period squared is equal to its mean sun distance."

Amanda looked around and said, "What does this all mean?" and she had a very confused look on her face. Charlie sat up straight, cleared his throat and began to speak. Pete looked at Charlie and gave him his most evil glare. He thinks he's so smart, he's showing off for her.

Charlie was explaining the laws in very simple terms. "When I was at space camp, we made ellipses, and measured the eccentricity, which is how flat or circular the orbital path actually is, mathematically speaking." He threw his head back slightly, and then began again avoiding Pete's eyes.

"His second law is pretty easy to understand, we tested it by counting squares inside an ellipse drawn for us on a grid. There were twelve pieces, like a pizza, and every piece had the same number of squares, no matter how the piece looked." Both of the girls looked rather confused, and so he tried again. "It's like pulling your gum out of your mouth into a long wire, and then spinning it around your finger." At that Pete let out a loud sigh and put his head in his hands. Charlie ignored him and said, "As you spin the gum around your finger it goes faster as it gets closer to your finger." He looked around at them, "It means that when the earth is closer to the sun, it travels faster in its orbit, and when it's farther away, it travels slower, and that the earth covers the same amount of space in the same amount of time." This everyone seemed to get.

"Now his third law is more complicated, so to make it real easy I'll just say that the closer a planet is to the sun, the faster it travels in its orbit. The farther a planet is from the sun, the slower it will travel in its orbit," said Charlie. He smiled at his group quite proud that he could explain all three laws of planetary motion.

Pete pulled his head out of his arms and again glared at Charlie. He realized that if he said anything, he would look spiteful, and possibly stupid, so he chose just to stare.

Amanda looked between the two boys and chose to aim her attention at Charlie, since he was trying to help them get farther in their endeavor. "So what do you think we should do for our presentation? I'm not sure how to present this to the class."

Just as Charlie was going to give his newly formed idea, Elizabeth ventured, "I think we should teach it, you know, have the class do activities and we be the teachers." She too was now sitting straight up and her head took a slight movement backward.

Charlie gave her a hard glance and said, "That's what I was thinking we should do." "Good," said Amanda and she looked at Pete for his agreement. Pete shook his head slightly and said, "I'll look up the third law, if you don't mind." Everyone agreed.

Elizabeth chimed in before anyone else and said, "Can I have the first law, I really want to make the ellipse things?"

"Yeah, okay," said Charlie. "I'll take the second law and find that activity we did in space camp." Everyone nodded and then Amanda looked around confused as to what she might do. It was Pete who spoke. He extended his hand towards her and drummed on the desk as he spoke, "Amanda, how about you find information on Kepler the person. When he lived? What he did? Who he hung out with? You know personal stuff."

"That is a really good idea," Charlie said looking from Pete to Amanda. He turned so that he was facing just her and could not see the other two faces. "That would give the class a chance to understand what was happening at the time of his discoveries."

Suddenly, Pete's idea had now seemed very much like Charlie's idea, and he was taking credit for it. Pete was seething mad, and clenched his fist and hit the table.

Amanda turned suddenly to face Pete. Elizabeth said, "I think that is a really good idea," and her gaze rested on Pete. "We all do research tonight and this weekend, and then we start putting it together in class." She smiled after she spoke, and it was just that moment the teacher asked the class to take out their English books.

Lunch was a release for Pete since he didn't have to compete with Charlie. When Davy's class came to the cafeteria, Pete was already halfway done with his lunch. "What took you guys so long to get down here?" Pete said, annoyed with Davy being so late two days in a row.

Davy sat down slowly at the table already filled with boys and said, "Gee Pete, I dunno, we just do a lot of stuff and I almost forget about lunch altogether."

Pete, realizing what he just said, replied, "Sorry, I'm just so mad at my teacher."

"Why are you mad at your teacher?" said Davy

"She moved our desks. She put them in groups of four and I sit with Amanda and Charlie," said Pete. "I want to sit by Amanda, but that Charlie is such a jerk, he's always trying to look so smart, and get Amanda's attention. If I tell the teacher I don't like Charlie and want to move, then I can't sit by Amanda. See how things are so messed up? And its only the second day of school." He picked up his sandwich and chomped down in it, chewing like a mad man.

Davy watched and ate his sandwich. "Yeah, all messed up," he said, but messed up for him was different. He was wondering how his best friend could find a girl so important. More important than him, it seemed. For the first time, he ate his lunch without talking.

CHAPTER SEVEN

GRANDMA'S TREASURE

Charlie's mom picked him up after school on Tuesdays so he could make his music lessons by 3:30 p.m. On his way down the Main Street, he saw his friend Billy Martin hanging around with a group of older boys who were in the high school. He looked like one of them because he was physically bigger. Charlie turned his head to get a good look at the kids he was standing with, and saw Mike Sullivan sitting on the bench in the middle of all the boys.

Charlie settled back in his seat and groaned. He couldn't stand Mike Sullivan, not many people could. He had the dubious distinction of being the only kid expelled from the eight grade, and he had to go to an alternative school for 90 days. He was now in the high school, but everyone knew that he smoked, drank, cussed, and probably did drugs. Most kids avoided him, except for those who thought he was cool and wanted to be like him.

What was that big, dumb Billy doing hanging out with him anyway? Charlie knew Billy's mom wouldn't like him hanging around Mike, and would have a cow if he ever got caught anywhere near him.

He looked out the window and a smile grew over his face. He was so happy with himself for making Pete look stupid, or least not as smart as him. It was bad enough that the two of them rode the same bus. He decided that no matter how hard he had to work, it was going to be him that was friends with Amanda, and hopefully more than friends in the future. He liked that idea very much, more than friends, but first he had to eliminate the competition. Pete with his good looks, and what luck, of all the kids he had to sit next to, it had to be him. It was going to take some effort, but he was going to get all of Amanda's attention.

Pete was very much on the same quest, to impress this new girl. To do this, he imagined that he would have to bone up on this Kepler guy so he could look just as smart as Charlie.

When he arrived at home, he went right to the refrigerator, he was starving. He opened the door and stood in front of it, like all kids his age. "Don't stand in front of the ice-box with the door open. Get what you need and get out," yelled is grandmother from the mud room. How she knew what he was doing was uncanny. Pete grabbed a green apple and some juice. He cut the apple up into four parts and gave Madelyn and Ashlyn each a piece and then poured them all a glass of juice. He patted the head of the closest girl, and left the room.

He moved quickly to the computer and started a search for Johannes Kepler. As he was reading through the laws again, something strange caught his eye. It was a sun clock, and it was very much like the object he found in the basement, that now was tucked under his bed up in his room. He looked closer and saw that it had a few more pieces attached. The photo showed an arm that came up and sat on the top near the big N, and a string that extended from the top of the piece to the S at the bottom. The quotation read, "In the northern hemisphere, the sun will appear in the southern sky. To correctly tell time by the sun clock, align it to the North Pole, and the string will cast a shadow on the number. The time is read by looking at the shadow."

Pete ran upstairs and carefully extracted the object from under his bed. He looked for a place where a bar of some type could be attached, but didn't see one on the front. He carefully rotated the object to the back and there he saw a place where a little small plate was located. He tried pushing it in either direction, but it wouldn't budge. "Great, I found a broken one," he said and kept trying to push it in either direction.

"Pete, Peter. Could you come down and take out the garbage before your granddad does it?" yelled his mother.

"In a minute" he said, and kept trying to move the metal plate in the back of the object by sliding it to either direction.

"Peter! NOW!" yelled his mother louder.

"Okay, Okay, I'm coming," yelled Peter and tossed the sun clock onto his bed.

When Pete had left his room, two sets of little eyes peered around the corner. "See, its right there on his bed," said a little voice. "What is it?" said the other. "I dunno, but it looks cool," and with that, two little sets of bare feet ran into Pete's room. Madelyn reached the bed first and picked up the sun clock and began to run her fingers through all the groves. Ashlyn reached for it, but Madelyn pulled it away. "My turn," she said, and held it close to her nose to get a better look.

Ashlyn stood with her small hands on her hips and leaned her head in to get a better look. She said, "What's that?" and pointed to the round glass, half globe that housed the compass needle. She reached over and pushed on the globe, and a metal bar shot out of the top. Both girls froze for a moment, and then Madelyn gave Ashlyn a sharp look and said, "I think you broke it." Madelyn started to wiggle the bar, and quickly it snapped into an upright position.

Again, the girls suddenly froze. "Now I think you really broke it," said the other. They were exchanging a worried glance between them when Pete suddenly entered the room.

"Hey, what are you doing with my stuff?" and that was when the girls turned to face Pete. The one holding the sun clock quickly shoved it behind her back to hide it. "Nothing," they said in unison. It was funny how they always seemed to say the same thing at about the same time.

"Let me see what you got there," said Pete, suddenly remembering what he left on his bed. He abruptly reached behind the one with her hands behind her back and pulled out the object.

"She did it," yelled Madelyn and pointed towards the other. "No it was her, she did it," yelled Ashlyn pushing her sisters pointing hand away, and then the yelling between them began, and the pointing, and the swinging of little hands.

"Hey, Stop It!" yelled Pete, and he stood between the two girls.

"Tell me who did this?" he said to them.

Madelyn turned to face him and said, "She pushed that thing," and she pointed to the clear globe housing the compass, "and that's when that thing popped out," and she pointed to the metal bar.

"Yeah, but you pushed it up and then it did that," and she showed how the metal bar was now fixed in an upright position over the N. Pete saw that at the end of the bar there was a small ring. He touched the ring and noticed it was attached to a piece of string. He slowly pulled it, and when he let go, it shot back inside the bar. He pulled it again.

"It goes there," a little finger pointed to the little hook below the S. "Where?" said Pete. The little girl pointed to a barely visible little hook shape that was part of the ray of sun just below the S. She pushed her little hand to the ring and carefully pulled it, and then holding the object to steady it with her other hand, she hooked it through. "There," she said and stepped back.

"Yeah, now what do you do with it?" said Madelyn. He looked up at the girls and said, "What are you doing in my room?" The two little girls looked at each other and ran out together. "Scram, and ask before you come in." Pete smiled, and closed his door. He didn't mind his sisters, but he was afraid that they could really break something.

Now the big question, how to use this contraption. He reached around and grabbed a t-shirt sitting on his bed. He carefully wrapped his shirt around the object, and tucked it under his arm. He went downstairs by the back stairwell, and sank into the chair by the computer. The object on his lap was hidden from view.

He typed in 'sun clock' and did a web search. He began opening web sites, one at a time until he saw a sun clock similar to the one in his lap. He began to read. "Place the sun clock so the compass is aligned north to south," he read out loud. "The string will cast a shadow, and the number on which the shadow is cast will reveal the time." He bent in close and looked carefully at the picture. He could see the sun clock sitting on the ground, and a shadow was cast between the 1 and 2. The caption below said 1:30 p.m. His opened the shirt and looked at the sun clock in his lap.

Before getting up and going outside, he looked around to see if he was going to run into his grandmother. He remembered that his granddad told him not to show her what he had. When he saw that she was nowhere around, he got up and went out the front door. He walked quickly and purposefully down the front porch steps not noticing the people sitting on the porch swing.

He found a sunny spot, and then held the compass in his hand so that the needle started to wobble and swing around. When the needle stopped moving and was set in its north-south position, he carefully set it down in the sun. He moved out of the way so his shadow was not on the clock, and he looked carefully to see where the shadow of the string fell. It was close to the 4. He rolled his hand over to look at his watch, and it read 4:55 p.m.

"What?" he looked again at the shadow on the sun clock and sure enough it was almost on the 4. "This stupid thing doesn't work," he said out loud.

"Let me see," said a voice from overhead. His grandmother was standing off to the side so her shadow was not on the sun clock.

"Yep, it's almost on the 4," and she shook her head. "It right," she said and looked at Pete.

"What does it say?" said a second voice coming from the porch swing. "Almost 4," said grandma back to Pops.

"Yeah, that's right," said Pops.

Pete stood up and looked at his watch. He pointed to his arm and said, "How can it be right? My watch says 4:59 p.m."

"Don't they teach you anything in that fancy school of yours?" said his grandmother. "The sun clock is telling you sun time. That is the time according to the sun."

"Well if this is sun time, then how come sun time is wrong?" said Pete. "My watch is telling me the real time, and it says that it's only 5:00, not 4:00."

Pete's grandfather got up off of the swing and looked sternly at Pete. "You will watch how you speak to your grandmother," he said. "If you would let her speak, she will tell you the big secret you can't figure out."

Pete dropped his gaze to his feet and mumbled, "I'm sorry Grandma."

Pete's grandmother moved in closer and put her hand on Pete's head. "So you found my sun clock. I love that thing. How did you figure out how it works?" And she bent down and picked it up off the sidewalk. She looked at Pete and said in a soft voice, "A sun clock tells you the true sun time. If you used this clock in Indiana, it would be correct."

"Why Indiana, why not here?" said Pete.

"Because my dear boy, Indiana is the only state that I know of that does not change their clocks to daylight savings time. You see, the rest of the country moved their clocks ahead, 'spring forward', and so the sun clock is off by one hour." She grinned at Pete, and handed the clock back.

Pete took the clock carefully from his grandmother's graceful hands and looked into her hazel eyes. "How do you know so much about this?" he said.

"Because I used to study Astronomy, and I hoped one day to be an Astronomer, but there was a war, and people had to do other things to get by, but I still remember," and she pointed to her head. "Let me show you how to take this apart," and she took the object and unhooked the string and let it gently spring back into the metal arm. The she wiggled the bar, and it shot upward into a flat position. Then she pushed in the globe, and the metal arm disappeared into the top of the sun clock. "I'm really impressed that you figured this out," and she handed it back to Pete.

Pete looked directly into her eyes, "Grandma, can I have this, I mean, I didn't know it was yours."

She nodded her head and said, "Do you promise to take good care of it?"

Pete put two finders up in a scout's honor signal and said, "Oh yeah, in fact I'm going to take it to school and show it to my class. Hey, can I show you something else?"

She looked at him and said, "What else do you have?"

"Wait right here," he turned and ran up the stairs. He stopped at the door and looked at his grandmother, "I'll be right back," and he was gone.

Pete's grandmother looked at the old man sitting on the porch, "You didn't tell me how excited he got when he showed you the sun clock."

He responded, "I didn't realize that he would act like this," and he lifted his hands with a palms up to show his confusion.

She was going to sit back on the porch, but Pete came storming out of the house so quickly, with something larger in his hands.

"Grandma, Grandma," he said panting, "Could you show me how to use this?" And he shoved the sexton into her hands.

She looked down at it and smiled. "Sure, just slow down a bit," and she patted him on the back. When she was sure he caught his breath, she said. "What you got here, my boy, is a sexton. This tool gives the altitude of the sun at anytime, but you must be very careful," and she gave him a long look filled with concern. "You could damage your eyes if you don't use this correctly."

Pete looked confused. "How?"

She began by showing him the mirror. "See this, a reflection of the sun will be shown here, and if you look directly at it, then you will hurt your eyes. "These," and she pointed to the two pieces of colored glass, "must be down," and she rotated them in front of the mirror, "so you don't hurt your eyes." She held the sexton up to her eye and looked off into the horizon. "First you need to line up the horizon like so, half of it goes off into the mirror," and she pointed one end toward the horizon. "Then you have to find the sun in the mirror, and adjust this arm," and as she spoke she adjusted the round wedge, "and there you see the horizon and the bright spot of the sun." She pulled it back and looked at the arm, "20 degrees, that's how far the sun is above the horizon right now. It's called the altitude of the sun. It's setting, and soon it will slide right off the horizon."

She looked at Pete who was mesmerized. "Did you get any of that?" She handed him the sexton, "Here you try and see if you can find that altitude."

Pete took the sexton and put it to his eye. His grandmother stepped behind him and whispered the instruction again in his ear as he went through the same steps. He pulled the instrument from his eye and looked at the calibrations, and there it was, 20 degrees.

"COOL!" he said, and repeated the steps again.

"Grandma, can I have this too?" he said looking up into her kind face. "I promise I'll take good care of it."

Again, she smiled at him. "What can I say?" And she hugged him.

Pete looked up and asked, "Grandma, what do you about Kepler? You know the guy who came up with the rules about how the planets move around."

"You mean the guy who came up with the Laws of Planetary Motion?"

Pete was so impressed, "Yeah, that's the guy."

She wrapped her hand around his shoulder and started to walk towards the house. "Oh, I have many books on him, let's go inside and I will show you."

"Hey grandma, what else do you have?" said Pete, "I mean gadgets like this one?"

"You would be surprised."

Pete liked the sound of that.

CHAPTER EIGHT

PETE EXCEEDS ALL EXPECTATION

Pete down-loaded diagrams and pictures that would allow him to teach Kepler's third law of planetary motion to the class. He made a clever power point presentation, and when he was done, he sat back very happy with himself. Amanda was definitely going to be impressed with him, especially when he demonstrated how to use the sun clock and the sexton.

He put the power point onto a disk, and placed it into his backpack. He wrapped the sun clock and the sexton in a beach towel, and put them into his back pack also. He was ready for school, and for the first time in a long time, he couldn't wait to get there.

On the ride to school, he told Amanda that he something really cool to show her, but she would have to wait until he could show the whole group. He didn't want to break the items getting them out on the bus.

Once inside, Pete waited until everyone in his group was seated and they were waiting for the class to quiet down and get started. He was about ready to explode with his news, when Charlie said, "My stepdad knows this guy who can bring in a telescope to show the class, if we want to add it to our presentation."

Pete was clearly taken off guard, "What do we need a telescope for, we're talking about planetary motion?"

"Duh, he could show us how to find the planets," said Charlie. He was looking directly at Pete, hoping to stare him down.

"You can't see planets during the day. What are we going to have everyone do, come back at night?" retorted Pete.

"He could talk about the planets, and when you can see them," replied Charlie.

"I have something better to use, and you can use it during the day, and it shows you more," said Pete. He reached into his backpack and pulled out the beach towel.

Upon seeing the towel, Charlie scowled and said, "Oh, you're going to show us your swim suit, or maybe just your goggles," He started to force a laugh but stopped himself when he saw Pete's treasures.

"These," Pete started, "are the tools of the Astronomer. They may not be a fancy telescope," and scowled back at Charlie, "but they will tell you all about time and the location of the sun." Looking directly into Charlie's eyes he finished, "during the day."

"This one," and he picked up the larger object, "is a sexton, and it will tell you the altitude of the sun in the sky at any given time. This one," he picked up the other object, "tells the time during the day. It's a sun clock." He pushed on the globe and out popped the arm, he pulled it into the upright position and took the string out and looped the end around the hook at the bottom. "You have to align the compass to north-south, and when the sun hits the string, it casts a shadow, and the shadow tells the time. Oh, but there's one thing that interferes with the correct time," and he looked around the group, "do you know what it is?" They all just stared at him. "It's daylight savings time. It makes the sun clock off by one hour."

Pete was about to demonstrate how to use the sexton when the teacher walked over to his table. "What is all this?" said Ms. Nagel.

Pete looked up, surprised by his teacher. "These are some old things I found in the basement. My grandma let me have them. She used to be an Astronomer, and she showed me how to use them. This one is a sun clock," and he held it up for her to see, "and this one is a sexton." He handed it to her.

She turned it over in her hands and said, "Could you teach the class how to use this?"

"Sure," said Pete. Out of the corner of his eye, he saw Amanda looking at him. His faced turned red and he looked up at the teacher, "sure, but could we make it part of our presentation?"

Ms. Nagel nodded and said, "That would be great. We'll let your group go first."

Then she went to the front of the class and announced that the presentations would start next Monday, and Pete's group has agreed to go first.

Elizabeth gave Pete a stern look. "I'm not sure I'll be ready by Monday," she said.

Charlie sat up, "I'm all ready to go," he said.

"So am I" said Pete.

"Well, I guess we better get to work on it," Amanda said to Elizabeth.

"I could help you if you would like," said Pete looking in Amanda's direction.

Amanda was just about to accept his invitation when Elizabeth said abruptly, "I would like some help."

Pete glanced over at her, and back to Amanda.

"If I am going to get this dumb thing done, I could really use some help," repeated Elizabeth.

Pete looked back at her, trying to tell her with his eyes that the invitation was not extended to her, and then he said, "Okay, maybe after school we could all get together at the library and work your part out together." He looked back to Amanda and Charlie was whispering in her ear. She shook her head and he knew that Charlie was trying to get her to agree to let him help her. He had to get her to agree to come to the library, or he would lose her to Charlie.

"Can you come after school to the library Amanda?" said Pete.

Amanda smiled at both of the boys, "Charlie would that be Okay with you?" Charlie agreed, and said, "Can I see those things you brought in?" He seemed very sincere, and Pete thought he was genuinely interested. He slid the whole towel across the table towards Charlie. Charlie picked them up and said, "So how do you do this?" Pete jumped in seizing his opportunity to look smart.

After school, Pete raced to the library and was waiting outside on the bench. He wanted to be the first to meet Amanda. While he was sitting on the bench, Mike Sullivan, and a group of older boys, walked by. "Look at the geek," said one of the boys in the group. Mike stopped and gave Pete a long look. "Aren't you the kid that plays with water balloons?" Pete glared back. He wasn't about to give this kid an answer, he was outnumbered and they were all bigger than him. "What's wrong kid, cat got your tongue?" said Mike, and the whole group started to laugh. "You got a secret stash of water balloons hiding there?" and he moved closer to where Pete was sitting. Pete was getting very uncomfortable when another kid he didn't know said, "Hey, aren't you the kid whose dad is in Iraq?" Mike turned to look at the big kid, and said, "What are you talking about?" The big kid spoke up again, "Yeah, I saw him on the news. This kid's dad reports the news from the middle of the fighting." "No foolin," said another kid from the group. "That's gotta suck," he said. "Leave him alone, he's probably freaked out enough already," said the big kid, and they started to walk off down the street.

Mike looked around, seeing the others leaving, and shot Pete another evil glance. Apparently he wasn't worth their time, but before he walked off, he spat on the ground in front of Pete, and gave him another dirty look.

Pete had been holding his breath and let it out all at once. Every town had its bad element, and Mike Sullivan was a really bad one.

Amanda arrived, putting Pete at ease. They sat outside and talked awhile as they waited for the other two to show up.

"I really think those things you brought to school are cool. Our presentation is going to be the best." She smiled and he noticed her lips framed her perfect teeth, and curled nicely into a winning smile. Pete could feel his face getting hot as he blushed. He looked away to see the crowd of boys just down the street. As soon as he saw Charlie and Elizabeth coming from the other way, he jumped up and said, "We really should go inside and get started." He didn't want the group of boys to come back and make a fool of him in front of Amanda.

In the library, the four found a computer terminal free and pulled four chairs around it. They looked up sites on Kepler, sun clocks, sextons, and laws that govern the planets' motions. They worked for two solid hours and everyone got along. Pete forgot that he was working with his natural enemy, Charlie. In fact, during class when Pete was showing Charlie the sun clock and sexton, he felt that Charlie was really interested, and not trying to show off for Amanda. After two hours, they were all ready to quit.

"I can finish on my own," said Elizabeth. She gathered up her papers and shuffled them into a stack, and shoved them into a folder. "I have to go, my mom wants me home by 6:00." She looked at the other three kids, "thanks for all the help" she said. "I never worked so hard with any other class mates in my life. I really feel good about this." She waved goodbye and left.

"We should probably all leave together," said Pete, remembering the kids who where gathered just down the block. After all, there was safety in numbers.

When they were just outside the library, Pete gave a quick look around, and noticed that the group of boys had gotten bigger. Pete looked off in the other direction, and said, "Anyone going down this way?" Amanda looked at him and said, "Wouldn't it be faster if we headed down Main Street and up Juniper?"

"Not really, I know a short cut that goes off this way," and he glanced nervously at her.

Charlie was now looking at the group of boys and saw Billy standing on the outside of the group. He too, turned to head off in the opposite direction when he heard his name being called from down the street. Charlie turned to see Billy waving for him to come over, and so Charlie looked at the small group and said, "I guess I'll see you later." He reluctantly turned and headed off in the direction of the group of boys.

Pete took this opportunity to start off quickly in the other direction. "Are you coming?" he said, and waited briefly before picking up the pace again.

"Pete, who are all those boys?" asked Amanda as she walked quickly to catch up with him.

"Oh, just some kids who hang out," said Pete without even looking back in that direction.

"It looks like Charlie knows them." She turned to glance at Charlie over her shoulder, and Pete took her hand. "Really, we should go," and he drug her down to the next block and rounded the corner. They walked two blocks out of their way, and after finally reaching Amanda's street, Pete stopped to say goodbye.

"Pete, we seemed to have walked a bit farther using your short cut," said Amanda. "Who were those boys anyway?"

Pete looked at his shoes and spoke, "they can be really mean, and I didn't want them to say something nasty to you." Actually, he was more afraid they would say something nasty about him, and he couldn't afford to look like a coward in front of her.

He looked her in the eyes and said, "We should be ready for a kick-butt presentation on Monday," and he smiled.

"You bet," she said, and smiled back. "I'll see you later," she turned to walk down her street.

He watched her walk away. She spun around once and waved, and then ran off towards her house. Pete was thankful she didn't make too big of a deal walking farther to get home.

Charlie arrived at the group of boys and Billy was saying something about him having a lot of money and being really cool. "We hear you're loaded boy, you want to share some of that dough?" said Mike Sullivan.

"My parents are loaded, not me, they hardly ever give me any money," said Charlie.

"Yeah, but he's still really cool," said Billy.

"You like to hang out with that fag-boy Pete?" said Mike.

"Me, uh, not really, we had to do some kind of assignment as a group, that's all," he lied.

"Who was that girl with you, she nice?" asked the big kid in the group.

"She's just a girl, no one really," said Charlie. He didn't care if they asked about him, but he didn't want them asking about Amanda. She was just a kid, and they need not be concerned with her.

"Hey, I got to scoot on home," said Charlie as he backed away from the group.

"Not yet," said Billy, "I want you to hang out with us awhile."

"Sorry, can't," he said, and quickly turned to leave.

"Hey, find out about that girl, the one with the red hair. I like red hair," laughed Mike.

Charlie cringed and made a face, then he turned and said, "Sure," and in almost a run, he went off down the street in the opposite direction glad to be away from that crowd.

Charlie was over prepared for Monday. He had prepared material for each part of the presentation, just in case one of his classmates forgot theirs, or didn't have sufficient information. He put on his nicest pair of dress pants and a button down shirt which he buttoned all the way to the top. He borrowed some of his mom's gel and swept his thick blonde hair over his head. He seriously thought about wearing a tie, he had a nice one and knew how to tie it, but didn't want to go overboard. What if Pete said something? And worse, what if Amanda thought he was showing off.

Pete had similar ideas about the day. He put on his best pair of slacks, a nice short sleeve button down shirt, and combed his wet hair across his head. He loaded all of his extra stuff carefully in his back pack, checked to see he had his disk, and then waited for his little sisters. This was going to be the best presentation ever.

From the moment both boys arrived at school, they received stares from the other kids in the class. Pete looked at Charlie, and Charlie returned the look. It was no big secret what each one was up to.

Amanda was very pleased to see her team members prepared and well-dressed for the event. She was wearing a plaid summer dress with a delicate necklace and matching earrings. She pulled her hair back, and let little curly tendrils fall around her face and neck. She was the prize to be won, and the boys were in severe competition.

Science wasn't until after math, and math seemed to drag on and on. When Ms. Nagel told the class to put away their math books, the butterflies began. Pete wasn't afraid of speaking in front of the class, he was afraid of not impressing Amanda. He needed to prepare mentally, and so he raised his hand and asked if he could go get a drink of water in the hallway.

He walked quickly to the drinking fountain and got a long drink. He was bent slightly over, letting the cold water hit lips when he heard the voice behind him.

"Are you almost done?" asked Charlie. It appeared that he, too, needed a mental minute.

Pete turned and looked at him. Charlie really wasn't such a bad guy. He had come to the library to help the girls prepare, and he seemed genuine about looking at his grandmother's objects. "I'm done," and he stepped away. He wiped his mouth with his hand, and then out of the blue, he didn't even know he was going to say anything, he blurted out, "I think we're going to do great today, I have confidence in our group."

Charlie took a long drink, and then stood up and faced Pete, "me too," he responded sincerely, and together they headed back to the classroom.

Amanda started off with background information about Johannes Kepler, his life in general, his work built upon other Astronomers, and his Laws of Planetary Motion.

Elizabeth stepped up to the plate and had a large piece of card board with a clean piece of white paper attached. She made two small "x's" and explained that these were the reference points of an ellipse, "they are called the foci," she said, and then place a strait pin in each of the x's. Then she took a piece of string, tied it into a loop and placed it around the pins. Taking her black marker, she put it into the loop, pulled it taught, and started to make what looked like an oblong circle. It turned out very nice. Then she stood up straight and said very clearly, "this is an ellipse, all planets follow this type of path, and the sun is located at one of the foci. Some planets have more round orbits, other have more oblong orbits. Pluto's orbit is so oblong, that it actually goes into the orbit of Neptune." She smiled and concluded, "Any questions?"

A hand went up immediately. A little girl with blonde hair asked, "When we are closer to the sun, or the foci, is that when we have our summer?" Elizabeth looked at her group and was hesitant to answer, when Charlie stepped in, "no, the reason we have seasons is because the earth is tilted on its axis and always points in the same direction. The earth moves around the Sun, and the amount of sunlight changes as we move."

The same hand went up, "well it seems that when we are closer it should be hotter."

Charlie replied again, "it would seem like that, but when we are closer to the sun, we have less sunlight in the northern hemisphere. This means less direct sun, shorter days, and we have winter."

Ms. Nagel interrupted, "Did you all hear what he said. Some people may think that distance plays a roll in the seasons. It does not."

"As a matter of fact, we are closest to the sun in January, on or about the 6th, and this is when we usually have some of our coldest weather." Charlie looked around, and said, "Any more questions?"

He picked up a stack of papers and passed them out. When he was done, he started talking, "Kepler's second law of planetary motion says that a planet will sweep an equal area in equal time." He then asked the members at the first group of tables to count the squares in the first slice of what looked like an oblong pie on top of a grid. He asked the second group to count the second group of squares in the second piece of pie, and so on. When they were done, he went to the board and asked for the results. He wrote down group one, and their results were 16 squares, groups two reported 16, and so did every group.

Charlie said to the class very loudly, "notice that each piece of pie is different, and that although they are different, they have the same amount of squares, therefore, a planet will sweep equal area in equal time." Then he handed out protractors and asked each group to measure in degrees of an angle, how far it was from the start of the piece of pie to the end. He went back to the board and recorded the data. This time, the numbers were different. Again, he stood straight up and loudly said, "this data indicates that in order to move the same amount of distance, a planet must speed up as it moves closer to the sun, and slow down the farther it is from the sun." He looked around, "any questions."

No one seemed to move. Charlie stood still looking at the class, and then he elbowed Pete to let him know it was time for him to go. Pete, startled by the push of Charlie's elbow, looked around and said, "Yeah." He went over to the board and drew a sun, and then what appeared to be circles around the sun. He stood back and began, "Kepler's third law states that the planet's orbital period, that is the time it takes to move completely around the sun, squared, is equal to the distance in astronomical units cubed." He looked at the class and could tell that what he just told them was not a language they understood.

He pointed at the Earth's and spoke, "the distance from the sun to the Earth is said to be 1 astronomical unit. Scientist use different units to measure in space, because space is so big, and since we are on the Earth, we use it as our base of reference. A planet like Mercury or Venus would have distance less than one astronomical unit." He pointed to Mars and said, "Mar's distance in astronomical units is 1.52. To cube this number, you just multiply it by the same number, 1.52, three times." He wrote 1.52 x 1.52 x 1.52 = 3.54 on the board. He turned and looked at the class, "so we take the distance cubed, and the answer is 3.54," and he circled the number. "Now Kepler says that the distance cubed is equal to the orbital period squared." He pointed to the board again, "scientist give the Earth an orbital period of 1, again as a base of reference. If the Earth is 1, then compared to the Earth, Mars would have to have a greater orbital period. The orbital period of Mars is 1.88 compared to that of the Earth. Kepler says that we square this number, that just means we multiply it by itself," and he drew 1.88 x 1.88 = 3.54 on the board, and he circled the answer. "Now you can see that the numbers are the same," and he put the chalk down and smiled. The class still looked a bit unsure.

"While studying Kepler's third law of planetary motion, I found this really amazing thing, it stumped scientists for a long time, its called retrograde motion." He moved over to the computer sitting near the front of the classroom, and slid in his disk. He moved the monitor so everyone could see what he was doing. He booted up what looked like the sun and then two planets with orbits circling the sun. "This is the orbit of the earth," he pressed a key and both planets began to move, he pressed the arrow up and the whole diagram changed position so now instead of looking down on the movement, it was seen sideways. "This other planet is Mars, now watch as the Earth and Mars move close to each other." The two planets got close, and then it appeared that the outer planet, Mars, briefly went backwards and then forwards again.

"Let me show you what that looks like in the night sky," and he pressed another few keys, and the screen went dark blue with white spots everywhere. Then a larger red dot started move across the screen with a streak of red behind it. It stopped, went backwards making a loop, and then forwards again and left the screen. "Scientist call this retrograde motion. It's an apparent motion because Mars never really moves backwards in its orbit." He pressed more keys, and another planet, this one with rings, did the same thing, but the loop was smaller. "Saturn, Jupiter, and all the planets with orbits bigger than the Earth appear to move backwards and make a loop, and then move forwards again." He pressed another key, and anther planet moved on the screen, and did the same thing.

He pressed more keys and the screen went back to the original one with the sun in the middle and the two planets moving in different orbits, at different speed around the sun. "Notice that the planet closest to the sun moves faster that the outer ones." He pointed to the planet closer to the sun moving faster, "Kepler's third law defines this, a planet's distance cubed is equal to its orbital period squared. So really what he is saying is that the closer the planet to the sun, the faster it moves in its orbit." He looked around and the students were nodding their heads.

"Where did you get such great software?" asked Ms. Nagel.

"I've been busy on the internet, but my grandmother who helped me, was once an astronomer," and he moved over to his desk and picked up the sun clock. "This," and he held it up for the class to see, "was hers, she gave it to me. It's a sun clock, and I would like to show you how to use it." He set it up and put it on the table in the middle of the class. "You align the compass north-south, and then the sun will cause this piece of string to cast a shadow. Where the shadow falls on the numbers, that's the time, but there is one problem."

"Oh, what is that?" said Ms. Nagel.

"During the summer when we use daylight savings time, the clock is off by one hour," said Pete.

He turned and went to his desk and picked up the sexton. "Astronomers use this to get the altitude of the sun. That is the height of the sun in the sky." He put it to his eye and explained how it is used. When he was finished, he put it down in the center of the table next to the sun clock. "If you want to look at them you can, but please be careful," and he stepped back.

Ms. Nagel put her hands together and looked around "are there any questions for this group?" She went over and looked at the objects in the center of the room. "This group gave an outstanding presentation. I think we need to give them a round of applause. With that, the class burst into clapping, and Pete's face turned bright red. He went back to his seat and collapsed in his chair.

Charlie sat down and looked at him, and gave him a thumb's up. Pete grinned and motioned back at him. Amanda came over and patted Pete on his head and slid into her chair. Even Elizabeth said, "That was really great guys."

Ms. Nagel went to the front of the class, "We really learned a lot about the movement of the planets. We learned all three of Kepler's laws of planetary motion, retrograde motion, and the bonus of the sun clock and sexton. Thank you for a job well done." She motioned for the class to clap their hands again.

As the teacher moved away, Amanda gazed at both of her very smart, and helpful partners. She felt lucky to have two very good friends.

CHAPTER NINE

I PLEDGE ALLEGIANCE TO THE FRIENDSHIP...

Pete sat next to Amanda at lunch and she smiled having him at her table. The other girls whispered to each other and giggled while Pete ignored them and spoke only to Amanda. After they had rehashed the presentation, Pete looked at Amanda and asked,

"You want to hang out after school?"

"Where are you going to hang out?" said Charlie as he moved in across the table. The two girls that had been sitting there moved down suddenly as Charlie squeezed in.

Pete, aware the competition was back sat up straight and said, "well, I was thinking we could go to the Ice Cream Bar just inside The Square, and then we could go down to the pond in the center of The Square and goof around, you know catch frogs, get our feet muddy, just hang out."

Charlie, needing very much to be part of this conversation, said "Or, the new movie at the Queen Theatre is Superman Returns, have you seen it yet?" he said to Amanda.

"I saw that movie," said a boy from down the table. It was Davy, he had moved in when some of the girls moved off. "I think we should go down to the pond, the dragon flies are moving through and they're really cool, they land right on you."

Great, all we need is Billy and we'll have the whole barrel of monkeys, thought Pete.

"What ya doin sittin over here," said Billy as he moved in next to Charlie.

Pete, avoiding eye contact with Billy, turned his body so he was only facing Amanda. "I'll talk to you on the bus, I'm done and I'm going to go outside." He leaned in closer to her ear, "It's getting crowded in here." He got up and threw his trash away. Davy followed and they left together.

"What about a movie?" Charlie insisted.

She smiled and said, "I don't think my mother would let me go to the movies with a boy."

"What about a group of boys?" and he shook his head. "I mean a group of kids, girls too?"

Amanda got up, "I'll have to ask her." She picked up her trash and left the table.

Billy, hearing all this said, "Did you just asked that girl out on a date?"

Charlie chewed his lunch more vigorously, "I didn't ask her on a date."

"Why was Pete sitting by you?" asked Billy.

"Look, we all worked together on a project, and we were just talking about it. Pete is not such a bad guy," and he got up also. Billy sat there until he finished his lunch. He didn't like that Charlie and Pete were eating lunch together. He didn't like it at all.

Pete was on the bus first, and he saved the seat next to him for Amanda. When she got on, she hesitated briefly, and then slid next to him on the seat.

"Pete, can I tell you something and you not get mad?" asked Amanda.

Pete suddenly got very nervous, "sure."

"I felt a little weird at lunch today, you know with you, then Charlie, and then that big kid Billy Martin. I would like to hang out with you both, but I don't want to have to choose between the two of you," she said.

Pete signed in relief, "Amanda, all I want to do is hang out with you, and maybe Davy too. I want you to meet him. He's a really nice kid."

Amanda shook her head, "okay, where do you usually hang out?"

"How about we meet at the pond in the center of the square? It's really cool. There are a lot of different lily pads floating on the pond, and we try to catch frogs and crawdads. Wear old shoes," he said.

"Okay," she replied slowly. "I don't know about catching crawdads, but it sounds fun. I'll meet you after I eat dinner."

The bus stopped, and she got off. "See you later then," yelled Pete before the door could close.

Pete turned to see her walk down her street and from the seat behind him chirped a little voice into Pete's ear, "I learned a poem today."

Pete turned around to look into the face of his sister Madelyn, "You did, can you say it for me."

"No, but it's really cool," she said back.

Ashlyn piped in, "Yeah, really cool, it tells you where the witches stand."

"What kind of poem are they teaching you that has witches in it?" said Pete.

"We stand up and say it every morning," said Ashlyn.

"Yeah, the teacher says it really loud, and she covers her heart with her hand when she says it," said Madelyn. Suddenly the bus stopped and so did the conversation of the witches. Pete was slightly concerned about what his sister's were learning in school.

Pete ate his dinner with amazing speed and when his plate was empty, he asked, "Can I be excused?"

His grandmother, who had just sat down looked around the table and said, "Why Pete, I just sat down, we all just started eating. Why don't you tell us about your day."

Unable to concentrate on what happened in his day, he blurted out, "The presentation went really well, everyone learned about the laws of planets, and they all liked the sun clock and sexton. Now can I go?"

Pete's mom intervened, "Glad to hear you had a great presentation, it sounds like all went very well. What about you girls, did you learn anything today?"

The twins passed a glance between them, and then Madelyn said, "We learned a poem in school." Then Ashlyn interjected, "Yeah, and we say it everyday."

"Oh really," said Pete's mom. "Can you say it for me?"

Again, they exchanged that look, "I can't remember how it goes, but it tells you where the witches stand," said one of the girls.

"Yeah, everyday we say where the witches stand," said Madelyn.

"What kind of crazy school is this, teaching kindergarten children about witches," said Pete's mom in a voice that sounded like she was angry.

Pete thought about this for the second time. What did one of them say on the bus? The teacher puts her hand on her heart and says....The Pledge of Allegiance...let's see. I pledge allegiance to the flag....and to the Republic for which it stands....Pete started laughing.

"I don't know what's so funny?" said his mom.

Pete looked around at his grandparents and then his mother. "They're saying The Pledge of Allegiance, and the part where they say, and to the Republic, for which it stands," said Pete.

Suddenly his grandmother started to laugh, and so did his granddad. Pete's mom ran the words over in her mind and she too started to smile. Madelyn said excitedly, "Yeah, that's the poem that tells us where the witches stand." And the little girls giggled.

"I guess it's the same kind of school you went to when you were a kid," said Pete's granddad, and took a second helping of peas.

After dinner, Pete raced to the pond as fast as could in hopes of catching Amanda. He didn't want to get their and have her already come and gone. He chose a spot in a clearing where you could see the entire pond, and sat down under the big pin oak tree.

Charlie, over hearing the conversation at lunch, went to the Ice Cream Bar, and when he didn't find anyone, headed down to the pond in the center of the square. On his way he saw the same group of boys he avoided the week before. He ducked behind a parked car, and carefully maneuvered down the path so he wouldn't be seen. When he reached the edge of the pond, he saw Pete sitting under a tree. He stayed behind a bush, so he could get to Amanda as she went down the path before she ever made it to Pete.

Amanda, not aware of the group of boys near the center of town, proceeded right past without even a thought for her own safety.

Mike Sullivan and a few other boys looked up as she passed by on her way to the center square. "Hey little girl, slow down. Where you going?" Mike said to her before she walked too far past.

Amanda looked around and saw the group of boys. She turned and walked on without a reply. This made Mike get up and start to walk after her. "Who does she think she is? Does she think she is too good to say anything to us?" The other boys just watched as he left. One yelled after him, "She's just a kid. You want to get into trouble?" The words only trailed after him. The other boys were not motivated enough to get up.

Amanda glanced around and saw the bigger kid was walking quickly after her. She turned around and started to run as panic filled her brain. Unsure of where she was going, she ran down to the edge of the water, and just as she stopped to look for Pete, a hand pulled her by the hair backwards. Amanda screamed as she fell on her back. Pete saw everything from the other side of the pond and was on his feet running over to her. Charlie, who had been behind the bush, only heard her scream, and then he saw Pete running off. He quickly decided to run in the same direction.

Amanda was half lying on the ground staring up at Mike's angry face.

"When someone talks to you, the polite thing to do is acknowledge that person," he said and kicked her, not hard, but mean spirited, in the bottom. "So, little girl, you wanna tell me where you're going in such a hurry?"

Amanda started to cry, and said in a very small voice, "I'm not going anywhere. Could you leave me alone?"

"Leave you alone!" shouted Mike. "Leave me alone," he said again in a mocking voice. "What a little scardy..." and that was all he got out before Pete jumped on him from behind.

"She said leave her alone. Why can't you listen to her?" and he hit the bigger boy in the eye with his fist. The older kid threw him off and started to run head first into his stomach when he was hit off balance again by another boy.

Charlie hit him from the side and pushed him on to the ground. He kicked at his feet as he screamed, "leave her alone."

Both boys stood over the bigger boy lying on the ground with their fists clenched.

"Okay, okay, let me get up and I'll leave her alone," said Mike as he rolled over on his hands and knees to get to his feet. "What's so great about her anyway, she's just an ugly mut."

With that both boys knocked him down again and started pounding on his body everywhere they could. It was Amanda who yelled, "Stop! Stop! You'll hurt him!" Hearing her voice, they both stopped and got off of the boy who had his arms covering his face.

Pete, with barely the breath to speak, said, "You ever touch her," and then Charlie finished the statement, "You even look at her, and I'll kill you." He looked at Pete and said, "We'll kill you," and Pete nodded with agreement.

The older boy got up and ran off up the trail. Charlie stuck out his hand toward Pete and he took it and shook. They looked at each other and then towards Amanda. "Are you okay?" said Pete. "I wish I would have gotten here earlier." Charlie nodded in agreement again, "me too," he said.

They walked off down the trail and together, pledging a silent allegiance to Amanda on that day.

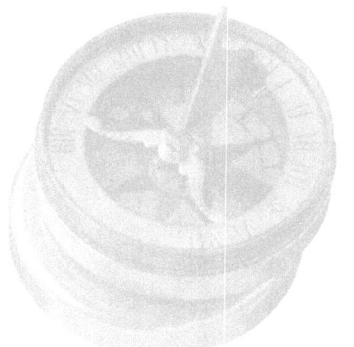

Chapter Ten

A Strange Phone Call

The phone picked up on the other end. "Hello, this is Ms. Nagel returning your phone call." A pause, and then a strong male voice said, "Thank you for getting back to me so quickly. I was wondering if you did what I asked at the beginning of school this year."

Ms. Nagel, "Yes, I put Pete and Charlie at the same table. At first I was reluctant to do so, seeing that they are both very strong role models. I was concerned that there would be a struggle for control, but they really have been getting along nicely. In fact, they compliment each other so well, they eclipse the rest of the class. It really isn't fair to the rest of the class to have them..."

"Sorry to interrupt," the male voice said very gruffly, "I'm glad to hear they get along so well. Will you be able to keep them together for the rest of the year?"

Ms. Nagel, "Well, I was really thinking I should split them up because I have other tables that are struggling and need a strong role model at those tables."

The gentleman's voice was a bit insistent, "So you're telling me that you have to move them apart because of the other students. If they are getting along and working well together, I don't see why they can't stay together."

Ms. Nagel responded, "I guess they could stay together, its just it would be more beneficial to the other students who are struggling."

"I don't care about the other students. The other students are not my concern. I care about my child, and I want him to stay in the group where he is, if it is possible," insisted the male voice.

Ms. Nagel, always anxious to keep a parent happy, agreed reluctantly. "I'll keep them together as long as they behave."

"What about the girl?"

Ms. Nagel sighed, "What girl are you talking about?"

"I believe her name is Amanda, is she going to be able to stay with the boys?" said the gentleman.

"She seems to get along with them very well, I really have no reason to move her away from them. I think if I move her, then they probably wouldn't get along at all. She seems to be the very glue that has bonded those two boys."

"Good, good. Keep them together as long as you can. Could you call me and keep me updated on how they are doing, maybe once a month?"

Ms. Nagel said, "I guess I could. Let me write your number down and I will try to get back to you if there is ever a problem."

"Very well, I look forward to hearing from you," and he hung up.

Ms. Nagel looked at the phone and frowned. She hated when parents tried to tell her how to run her classroom.

Chapter Eleven

Davy's Turn

Davy was making his way down the street to Pete's house on this bright Saturday morning. He hadn't seen much of Pete lately. He was wondering if Pete was going to hang out with him, or that girl. Just what was it about that girl that made Pete act so funny, and why was Pete now talking to Charlie and acting like they were friends? School had started four weeks ago, and already things were messed up.

He kicked up the dust as he strolled into Pete's front yard. Pete was so lucky to live in such a house, well, mansion really. Not many houses in this town had three stories and a basement, and two sets of stairs for each floor. The house did need some updates and repairs, but still it was a great house.

He walked up the front steps to the huge porch that wrapped all the way around the house, and knocked on the door. He wasn't sure if he was happy to be here, or sad because he could be disappointed that Pete would be busy with someone else, like he had last weekend.

Pete's grandmother answered the door. She was a graceful and pretty lady, thin and kept her hair up neatly on the back of her head. She always smelled nice, like flowers, and she always wore a smile. "Hi there Davy, come on in," she said as she opened the door wide enough for him to slide through. "Haven't seen you for awhile, been busy?" she asked. "Let me call Peter for you."

She turned and yelled, "Peter, Peter darling, your friend Davy is here for you." She smiled that friendly smile at him and said, "Why don't you come into the kitchen and have a muffin. Pete's mother just took them out of the oven and they smell wonderful." She put her arm around his shoulder and led him to the kitchen.

Davy took a muffin, said thank you and smiled back. He was happy to be here, at least at that moment.

Pete came running down the back stairs, "Mom do you know where my brown shoes are?" he blurted out.

"Where you left them," she replied without even looking up.

Pete's grandmother turned with something in her hand, and there they were. She handed them to Pete with a smile.

"Thanks Grandma," said Pete, and took the shoes. Without putting them on, he went out to the back of the house on the porch. Davy followed him out eating the muffin, "thanks," he said as he left the house.

Pete sat dawn on the bench and began putting on his shoes.

"Hey Davy, I want you to meet someone," said Pete without looking up.

"Yeah. Who? Is it that girl?"

Pete stopped and looked at him. The tone of his voice gave it away. "Yeah," he said as he continued to stare at his friend.

"Why?" said Davy.

"Cause I think you would like her," replied Pete.

He put his other shoe on and then settled back in the chair. He continued to look at Davy and ventured, "She's not like other girls, I swear."

Davy shook his head, "Okay, I'll meet her, if you want." His voice lacked all enthusiasm. Davy had no other choice really. He would either suffer through this girl and have Pete as a friend, or have no friend at all.

"Davy, I really think you might like her," said Pete. "I really like her, and so does Charlie."

There it was, what was really bothering him, Charlie McMullin. If Pete became friends with him, then Davy would be out for sure. When did Pete start liking Charlie? Things were changing so fast. First this girl, then Charlie. What next, Billy Martin?

"Are you and Charlie friends now?" asked Davy.

Pete froze. The thought of him and Charlie being friends had never really crossed his mind. They were talking like friends, and they both hung around with Amanda, who was their friend, but was he friends with Charlie? "I dunno, why?" said Pete.

"Well, if you and Charlie become friends, what's going to happen with Billy?" worried Davy.

Again Pete stopped and pondered this conundrum. If he and Charlie were friends, how did Billy, and Davy fit in to the equation?

"You're my friend, and I will be friends with you no matter what happens with them," said Pete. He meant every word, but still he was getting worried.

Pete's grandmother walked out on the porch and touched Pete's shoulder gently. "I would like to show you something later, not now, but later. It's about Astronomy," and she winked at Pete.

"Why not now?" he said.

"Later, when the time is right," she said, and winked again. "Go on now, go play. There is so little time in life to just play." With that, she went back inside and let the door slam behind her.

"Let's go," said Pete, "Amanda is going to meet us by the pond." He got up, waited for Davy, and together they left the porch in the direction of the pond.

On the walk there, Pete was thinking of what would be the best way to manage this meeting. He wanted Davy to like Amanda, but he knew Davy wasn't very fond of girls. As they walked, Davy was quiet, which was not like Davy. Pete knew this was going to be hard, if not impossible. What if they had a miserable time and Amanda refused to ever hang out with Davy again? What would he do? Would he have to split his time between the two of them? Was Davy right now thinking of how awful it was that he insist we hang out with a girl?

Davy, in turn, was thinking how awful it was that Pete wanted to hang out with a girl. How could he do this to him? They were best friends, and now there was this girl. His insides were so upset that he couldn't even beep a single sound. He just walked along next to Pete staring at his old, beat up shoes. At least he would get his feet wet, but today with this girl thing, that might not even make him happy.

Amanda was already next to the pond looking into the water. She was talking to someone because as they got closer, they could hear her voice. When they rounded the bushes, they saw that it was Charlie.

"Oh great," groaned Davy under his breath.

"Hi Pete," said Amanda as she gazed in his direction.

"Hi guys," said Pete, "I hope you don't mind, I brought Davy, he's my friend," and he pointed to Davy.

Davy scratched his head and turned his body around to look at the other end of the pond. I hope you don't mind, rang through his head. "Hi," said Davy to the make-believe people on the opposite side.

Pete pushed Davy with his shoulder, and Davy turned to face the other two. "What are you looking at?" said Pete.

Amanda smiled briefly, and then pointed to the lily pad. "There's a frog just sitting on that lily pad, right there," and she tried to reach it with a stick.

"If you poke it, it will jump into the water," said Davy.

"Here, let me get it for you," said Pete, and he started walking slowly through the water. Davy just stood at the edge and watched as his best friend fetched a frog for a girl.

Charlie was crouched down at the water's edge ready to catch the frog if it came near him. The three of them looked really cozy at the side of the pond together. Davy shook his head and a tear came to his eye. He knew Pete was more popular, a better athlete, and smarter, but he didn't think he would lose him so soon. He looked at his feet, unable to look into the direction of the other kids.

A splash broke the water, and the three kids started to laugh. Amanda came out of the water dripping wet and laughing loudly. "What am I suppose to do now?" she said shaking her hands so the water would fly off. "That darn frog jumped on me. Did you see that frog?"

Pete was holding his stomach laughing and Charlie was rolling on the ground. "It landed right on your face," Pete said with difficulty. "I'm surprised it didn't pee on you," said Charlie.

"PEE ON ME!" shouted Amanda, "You mean they pee on you?"

Davy started to laugh. The thought of frog hopping on this girl's face and peeing on her was too much.

"Look, another frog," said Davy as he pointed to the edge of the pond, "and it's huge."

The laughing stopped as the other three kids moved closer to get a look. Sure enough, the biggest frog any of them had ever seen was hidden just below the surface. Only the eyes peered at them.

"He doesn't see me," said Davy, and he slowly crept closer. "Get ready for him to jump in your direction," he said.

Charlie moved to the edge of the water, and Pete, already in the pond put out his hands.

"Is everyone ready?" said Davy, and he slowly reached down with steady hands to scoop up the frog. Slowly, slowly, everyone watched with anticipation, and then he swooped in, and totally by surprise, he scooped up the huge frog and brought it back with him out of the water. "I got it!" he yelled. He fell to the shore and squeezed the squishy frog gently with his hands. It squirmed, and he had to struggle to keep hold of it. When he sat up, all three kids were gathered around him.

"Let's see the beast," shouted Amanda, filled with excitement.

Davy couldn't help but laugh as he produced the biggest frog he had ever held. "You want to hold it?" he said to the others.

"Yeah, gimme it," cried Pete. "NO!" exclaimed Charlie. "Let Amanda hold it." Everyone looked at Amanda.

"Okay," she said as she wiped her hands on her wet pants. She reached for it, and then pulled her hands back quickly, "Will it pee on me?" she said. The boys roared with laughter. "It might," said Pete, "But its good luck if a frog pees on you."

"Really, I think it is good luck," said Charlie.

"Okay, let me have it," she said, and again reached for the frog.

"Have you ever held a frog before?" asked Davy.

Again, Amanda's hand stopped suddenly, and she wiped them on her pants.

"Not really," she said. "Is there a special trick to holding frogs?" she said to Davy and looked intently into the frog's little face. The frog blinked his eyes and croaked. "Maybe he doesn't like to be held," she said.

Davy giggled at her, "No, they really don't mind, just make sure you get the back legs, or it will jump right out of your hands."

"Are you ready? Here, let me hand him to you," and with that, Davy stuck out the frog for her to grab.

Amanda, unsure about holding this animal, looked around at the other boys and said, "Okay, now, here we go." Reluctantly, she put out her hands, pulled them back, and then put them out again. Davy pushed the large frog in her direction, and their hands met. Amanda put her hands around Davy's. Her eyes became very large, and her mouth opened, but nothing came out. Davy released the frog, and let his hands slip away. Amanda had the whole frog, in its entirety, in her hands. She stood completely still, and let out a slow breath.

"I have it! Look!" she shouted, and then it happened. The frog made a huge leap with its back legs, and Amanda, taken by surprise, dropped the frog and fell backward on her butt. "Ahhh!" she yelled, as the frog jumped on her chest. "Ahhh!" she continued to yell and fell on her back. The frog took the opportunity and jumped again. This time it landed on her hair, and Amanda screamed again. "Get it off! Get it off now!" Harried by the motion, the frog made one huge leap, and was in the water.

Pete and Charlie, shocked by the event, both jumped into the water to try to catch the frog. Davy was the one holding his stomach and laughing. Amanda looked at him and started laughing too. Pete and Charlie were in the water splashing.

Davy bent down and took Amanda's hand and said, "Are you okay?"

"I'm not hurt," she beamed at Davy.

Pete and Charlie suddenly stopped and both were looking at what was going on at the edge of the pond. It happened so quickly, they almost missed it.

Davy's face changed, and they knew the moment it happened. Amanda did 'it' to Davy. You could see it in his eyes. He too was hooked, reeled right in by the girl with the curly, auburn hair.

Amanda stood up with the help of Davy and said, "Is there somewhere we could go that's not so, messy?"

Charlie was first to address the question, "Yeah, let's go back to my house and have something to eat."

Everyone seemed to agree, especially Amanda. They made their way back up the path and down the street towards Charlie's house. The small group had grown from three to four.

Pete had never been to Charlie's house, and he was impressed. Charlie's house was big like his, but the grounds were immaculate, the inside was very well kept and nicely decorated. His family had two maids and a cook. He had a younger brother and sister, and lived with his mom and stepdad. When they came into the very nice kitchen, Charlie asked the cook to prepare them some lunch. He asked what they would like to eat, and the cook became very busy with the order.

"Wow, your house is so nice," said Amanda.

"Thank you," replied Charlie, "but we've only been here two years."

"Where did you live before that?" asked Pete.

"Well, we used to live in Detroit, and then my grandmother became ill, and we moved into her house to help her."

"So, this is your grandma's house?" asked Pete.

"Yeah, my mom's family is loaded. My real dad died when I was three, and then my mom married Howard, my stepdad. They knew each other ever since they were kids. They all lived here when they were younger, and couldn't wait to move back," said Charlie. "In fact, my grandmother knows yours," he said to Pete. "I think they were friends or something when they were young."

Pete didn't know what to say. He didn't really know much about his grandparents, this town, or the house he lived in.

While they ate lunch, they listened to Charlie tell about the cemetery behind the little church up on the hill, the well house down by the creek, and the old barn on the Welston property. Pete could not believe he didn't know about any of these places.

"Can we go there?" asked Davy, who voluntarily became one of the group.

"Yeah, no problem," said Charlie. "How about we check out the well house after lunch?"

Everyone nodded in agreement. Pete could barely believe his day. He had his best friend, Davy, his new friends Charlie, and Amanda, and a cool adventure. The day couldn't get much better.

CHAPTER TWELVE

THE MIDNIGHT DANCE

Pete wasn't tired, but his mother told him to go up to bed. It was like this every night. He was ordered to his room before the nine-o-clock news. Pete thought it was because his father may be reporting from some dangerous place, and his mother was afraid it would bother him. He hadn't heard from his father for almost a month, and lately it didn't seem to bother him as much as in the past. Pete had become so distracted with his new friends, school, and Astronomy that he had hardly even thought about the danger his father was really in.

Pete just turned out his light and was heading to his bed when a flash of light from the window caught his eye. He moved to the window to look out, and he saw his grandfather with a flashlight shining it towards his window. He looked closer and saw his grandmother standing next to him. Pete opened his window just a couple of inches. He stuck his head out of the window and heard his grandfather whisper to him. "Peter, Peter, come out here."

"What's going on?" whispered Pete back.

"Shush, just come out here. Use the back stairs, and don't let your mother see you," said his grandfather.

Pete picked up his flip-flops and tip-toed down the back stairs. Once he was outside, he put on his flip-flops and followed the light to where his grandparents where standing.

"Come here, my boy, I want to show you something," said his grandmother. She put her arm around his shoulder and led him to the clearing in the yard. The old house sat near the top of a hill, and so from where they were standing, they could see most of the sky.

"My father always called this 'big-sky' right here where you are standing. When I was a young girl, about your age, my father brought me out here, and showed me what I am going to show you. I feel bad because I never showed your father, but I tried one night, and he just wasn't interested," she said. "So Peter, I am going to show you, so you can show your little sisters and later your own children."

Pete was confused. "What am I going to see?" he said.

"I am going to show you how to read the night sky," she replied. "First, look all around you," and she held her arms out to signal for him to look all around. "See all of this?" and she waved her arms. "This is your very own celestial sphere, over your head. It moves wherever you go, and it is always changing. What you have to do, is know where you are in your celestial sphere."

She took one step away and pointed her right arm directly over her head, straight up. "See this point, directly over my head?" Pete nodded, "yes." "This is my zenith, and if you point your arm directly over your head, you will find your zenith. Here, try it," she said, and walked over and gently moved his arm over his head. He looked up at a bright star in the middle of the sky, "Wow, is that the North Star?" he said.

Pete's grandfather chuckled. He heard him from the porch swing. "He's only laughing because it's the exact same thing he said when I showed him how to read the night sky. That is not Polaris, the North Star. If you were at the North Pole, then it would be, but you're not at the North Pole."

"Then what star is that?"

"Well, we will see in just a minute, first let's find ourselves in our sphere. To do this we must find our Cardinal Points," she said.

"What's a Cardinal Point?" said Pete.

"A Cardinal Point is a compass direction," she replied. "If one knows his Cardinal Points, then one has a baring on his life." She smiled at him, and realized that what she just said went right over his head. She took a deep breath and started, "It's pretty easy to determine your Cardinal Points because the sun gives us this clue everyday. It rises in the east and sets in the west." She turned towards the house and said, "The sun rises on that side of the house, and sets every night on this side," and she turned and pointed opposite the house. "So if I put east at my right hand, and west at my left, then I am looking to my north." She turned her body to face north as she spoke. "Now, the Cardinal Points also have numbers. North is 0, and 360. East is 90, south is 180, and west is 270. When we talk about the compass direction to find a stellar object, we call this our azimuth. For example, the sun rises a little different each day, but always in the east. So if I were going to look for the sun to rise, I would look for it in the east."

It was a little confusing, but Pete was definitely interested. "Okay, but its night, how is this going to help us during the night?" asked Pete.

"Good question," said Grandma. "Let us start by finding the North Star, Polaris," she said. She turned her body to the north and said, "Now, we have to find our altitude of our star. To do this we use our body like a human protractor." She put her right arm straight up over her head again. Pete watched as his grandmother did these strange moves, almost like a dance. She put her left arm straight out in front of her, then she spoke again. "My left arm is pointing towards the horizon, we call the place where the sky and the land meet, zero. My right arm is pointing straight up, so the angle between them is 90 degrees. Do you know about degrees of a circle?" she said. She should have asked him before when she was explaining azimuth, but he seemed to understand.

"Yeah, I know there are 360 degrees in a circle, and that 90 degrees makes a right angle," said Pete.

"Good, good," she said slowly, "Then this is going to be easy for you." She stood with her arms fixed as she continued. "We are at 39 degrees north latitude, so I need to bring my left arm up about 39 degrees, and there," she said. She was pointing to a fairly dim star. "That, I believe, is Polaris. Now to make sure, we find the pointer stars," she said. She moved her head as to get a better look, and then pointed and said, "There they are, and they are pointing to the same star."

"How do those stars fit in?" said Pete.

"You see the drinking gourd?" she asked.

"The what?"

"The Big Dipper, it is one of the easiest star clusters, or constellations we can see. See the four there, they make the cup of the dipper, and those that stretch outward, they are the handle. But, it is the two from the dipper's end that point straight towards the North Star. The North Star is not the brightest star in our sky, in fact it is rather dim, but its significance is that it is directly above the rotational North Pole, and so we can use it to find other stars, and constellations," she said.

"I see it, the North Star, and I do see the Big Dipper. Wow, what else can we see?" asked Pete.

"Let's find the circumpolar constellations. These are the constellations that we can see every night, they never set, and they move around the North Star. Once we become familiar with them, I will teach you the others," said Grandma.

She took a moment and rubbed her chin. "I haven't looked at the stars in so long," she said in a voice that was sad, yet knowing. "Let's see," she said, and pointed up to the sky. "You see the four stars, and then the one on top. See how they make a house?" she said and pointed to the northern sky. "That is King Cepheus. You have to imagine him sitting in a chair in those stars." She moved her head to the left and said, "See the big E? That is his Queen Cassiopeia. She is always opposite the big dipper which is really the constellation the Big Bear, and the Little Dipper is part of the Little Bear. Can you see the stars that wrap around the little dipper?" She was pointing to the stars barely visible. "In most places you can't even see those stars because of too much background light, but that is Draco, the Dragon. When we go inside, I'll let you see my map of the sky, and then you can see the outline of the stars and the objects they are said to make."

"Who came up with the constellations?" asked Pete.

"It was the ancient Greeks who first looked at the sky, and they did so with amazing accuracy. They made up stories about how the earth was made, and the characters were immortalized as constellations, and placed in the night sky. It really does help us find stars," she said.

She took Pete by the shoulder and led him to a bench, and sat down. "You see, the earth is always moving, so in just a little while, all the constellations we just found will be in a different place in the sky. They are the easy ones to find. Because the earth is also revolving around the sun, we see different constellations throughout the year. The ones we see tonight will be different from the ones we may see 6 months from now."

"You mean we won't be able to see the bigger dipper in 6 months?" said Pete.

"Oh no, we will see the big dipper every night, all night. The big dipper belongs to a group of stars that will never set. The North Pole points towards the North Star always, and so we will always see that star and the ones that move around it," she said.

It was slowly beginning to sink in. "Have you had enough for the night?" asked his grandmother.

"I want to look at them again and see if I can find them," said Pete.

His grandmother smiled, and said "that's my boy."

Together they stood up, and went back to the clearing. "Start with your arms like mine," she said and again she put one out to the north directly to the horizon, and the other over her head. Pete did the same to imitate what he thought looked like a dance. They found Polaris, and then he found the other constellations they found before.

When they were done, his grandma said, "So what do you think about the stars now?"

"I really like looking at them when I know what I'm looking at," said Pete. "Can you show me more?"

"Sure," she said and turned to face the man sitting on the front porch. "Dear, where is that Star Finder you brought out?"

His head suddenly bobbed up to reveal his recent slumber. "What did you say dear?"

"I said, could you find that Star Finder you were playing with earlier," she said in a frustrated tone.

"Yeah, where is that thing, it was right here, it must have dropped," and he reached around and the swing came out from underneath him as he reached down to pick up the Star Finder. They heard a loud thud as his butt hit the floor.

"Good gracious dear, what have you done now!" cried grandma as she ascended the stairs.

"Woman I'm fine," he said as he reached for their hands to help him get up. Together, Pete and his grandmother pulled him off the floor.

"Are you okay Grandpa?" said Pete.

"Oh, I don't think any thing is broken," he said as he brushed off his pants. "Here is your star thing," and handed her the Star Finder.

She took it from him and gave him a weary look. "You sure you're Okay?"

"I'm fine, I'm just fine," he said and went and sat on a rocking chair this time.

Pete's grandmother moved over to the window which was giving off light from a lamp inside the house. She sat on a small bench on the porch, and motioned for him to join her. "This is an Astronomer's tool," she said. "It's really quite ingenious."

Pete looked at it confused. Right away he pointed out that west and east were opposite. "That's right," said his grandmother, "you're very observant. This is made so that you can dial up the day and time you want to observe," and while she was explaining, she was showing him how to dial up the day, and then she was pointing at the time. "This is made so that when you look up into the sky, you put this directly over your head, and can compare the real sky to the map in your hand."

"What is that piece of string that goes down the middle?" Pete asked and pointed to the string.

"That is your celestial meridian. It divides the sky into two halves, helping you find the constellations easier. Here, this is an easy one to find." She stepped off the porch and back into the clearing. She put the Star Finder over her head and moved so that she was facing due south. The she pointed off towards the southeast. "Pete, come here and I will show you the constellation Sagittarius."

Pete walked over to her, and she moved the Star Finder over his head. "Take it my boy," she said, and handed it to Pete. He took it and then she pointed to what looked like a Tea-pot. "See those stars on the Star Finder," she said. Pete looked at the tea-pot again, and then she said, "Look over to the south east," and she pointed to the group of stars near the horizon that looked the same, but bigger. "That is Sagittarius," she said.

Pete was amazed at how similar they looked. He looked at the Star Finder again, and out into the sky and recognized another constellation. As he stood there, he could find just about everything on the Star Finder.

"Wow Grandma, this thing is amazing."

She smiled a warm smile and said, "I'm glad you like it."

After about a half hour, he put the Star Finder down, and walked over to his grandmother. He put his arms around her waist and gave her a long hug. He looked into her delicate face and said, "I didn't know you knew so much. You're really smart for a grandma."

She smiled at him, "I'm glad you think so."

He released his hug and said, "Grandma, can I ask you a question?"

"Why sure. Come sit down. Is it about the stars?"

"Not really, I mean no. It's about Charlie's grandma."

"I'm not sure I know Charlie," she frowned.

"Well, Charlie is like a new friend," said Pete.

"Is he new at school?"

"No, he's been there for about two years, just about as long as I've been there, it's just that, we just now became friends."

"Oh, took awhile is what you mean," she laughed.

"Yeah, took awhile," said Pete. "Today, I was over at his house. It's the other really big house in town."

Grandma straightened her back and rubbed her chin. She seemed to rub her chin whenever she was trying to remember something. "Ah, yes, that would be the Fairchild Estate," she said.

"Are you sure? Charlie's last name is McMullin," said Pete.

She smiled, "If you're talking about the very large house with the most beautiful garden you have ever seen, then we are talking the Fairchild Estate. I knew Elise Fairchild, we went to school together. We were friends from the kindergarten to the ninth grade. That was when she stole my boyfriend." Pete's grandmother rubbed his head and messed up his hair. "You sure you want to hear the rumblings of an old woman?"

Pete's eyes were large, "yeah, I want to know."

She began again, "Elise was beautiful. She had blond-curly hair, big blue eyes, and a figure to die for. She could have any man she wanted. She flirted with every man, and often broke their hearts. I was no match for her, but we were friends because she knew she could outshine me, and I wasn't a threat. I had this real nice boyfriend. He was smart and handsome and devoted to me, until she decided she wanted him. I begged her not to do this, you know, flirt with him like she did. She said that she wasn't flirting, and that she couldn't help if men liked her more than me." Grandma looked down at her hands. "That little floozy took him from me. He was smitten with her and followed her around like a puppy, all men did. She kept him until she was sure she had ruined what we had, and then she cast him aside for her next victim. She got what she deserved."

Pete looked into her eyes, "What did she get?"

"Pregnant," said Grandma. "Funny thing was, she really didn't know who it belonged to. She had so many men knocking down her door that she just had to pick the one she thought was best, and marry him. He didn't even know if it was his or not. She picked this guy who was really handsome, but he was no good. He drank and smoked and beat her. After four or five years they divorced, and she left town. She came back years later with a grown son and daughter. The son went off to Kuwait and was killed, but the daughter was just like her. She tormented every young man in town, even your father for a while. He was too smart and distracted with his career to be interested in her very long. I think that if she could have gotten her claws into him, she would have tried to marry him. She married Donald McMullin's son and they moved away. I did hear she was back, but she's not a McMullin anymore." Grandma kept up with the gossip in town.

"I think her name is Jackson now," said Pete.

"That's right. She married Howard Jackson. Her husband sells used cars downtown, and he also does taxes," she said.

Pete added, "I really don't know that much. I do know he's friends with Billy Martin's father."

"Oh, that piece of trash," she said rather harshly. "He can go to hell in a handbag as far as I care."

Pete sat up, he never heard his grandmother talk like that before. "What happened to your boyfriend, the one she stole from you?"

"Oh that man," she reminisced. "Well, he was a lost soul. He moved away and worked for the railroad awhile, went to school and learned a trade, and then begged me for my forgiveness, and I married him."

"That guy was Grandpa?" Pete asked credulously.

"Yeah, I loved him and hated him. Someone once said that love and hate are the same emotion, it's just how you emphasize it that matters," she said and smiled as she gazed towards the porch. "I bet that old fool is asleep again." She got up and walked over to the foot of the stairs. "Yep, snoring away." She walked up the steps and nudged him with her hand on his knee. "Wake up Dear, the mosquitoes are going to carry you away," and then gently bent over, swept away the gray hair, and kissed him on the forehead.

Pete was amazed at how sweet they really were to each other. He wondered how long they had been in love. He was secretly glad she forgave him all those years ago.

She turned to Pete and said, "Darling would you help me get this old fool to bed?"

"Sure Grandma," and together they pulled him off the rocker and put an arm around each of their shoulders and walked him quietly inside the house. Pete caught the screen door before it could slam shut, as he held the Star Finder in the other hand. He hoped his grandmother didn't mind if it came to stay in his room for awhile.

CHAPTER THIRTEEN
BILLY'S TURN

Pete was trying to explain the way his grandmother showed him how to find the stars last night, but Davy was just shaking his head. "I don't understand," said Davy for the second time.

"I'm just going to have my grandma show you. I'll ask her if you can come over next time we go outside to look at the stars."

Davy and Pete were headed for the cemetery by the old church. They had all agreed to meet there today and snoop around. They found the trail that Charlie had described and were heading off for the church when Billy Martin stepped right in front of them on the path. "Where are you two goons going?" Surprising them, he pushed Davy hard in the chest. He fell backwards into Pete. Pete was almost knocked off of his feet, but recovered and caught Davy in his arms. Pete helped Davy stand back up, and said, "Get out of the way."

Billy put his hands on his hips and smiled. It was obvious to Pete that he was ready to cause trouble, when he heard another voice. It was Charlie, "Hey guys, it's up here just a little farther."

Billy turned and looked at Charlie, "Who are you talking to?" he said.

"All of you, come on now, let's go," said Charlie, and he turned and continued up the path.

Billy turned and walked up behind Charlie and muttered, "You gonna let those two bozos come with us?"

"I invited them, so shut up and come with us, or leave," replied Charlie.

Pete was impressed with Charlie for being so blunt with Billy. He and Davy exchanged glances, and followed farther behind on the trail.

The church was very old, and the doors were locked, but one of the windows was open. Davy looked in the window, "Does anyone want to go inside?"

"Let's wait until Amanda comes," said Charlie.

"Is she coming?" asked Pete.

"I called her this morning, and she said she would come up after church. She said it would be about 10:30, so I'm guessing she is going to be here soon."

Billy was standing away from the group watching them. He had his hands on his hips and his face was in a scowl. Charlie looked over at him, and looked away quickly. He was sorry he asked him to come given the way he was acting. He had made up his mind that if Billy didn't want to be his friend because he wanted to hang out with Pete, Amanda, and Davy, then that was just too bad. He didn't always like being with Billy.

Davy was snooping on the side of the church when his foot hit a log and he tumbled into the grass. Billy let out a loud snort, "you goof!" he yelled at Davy.

Davy got up and brushed himself off, and gave Billy a dirty look. He said nothing and disappeared behind the side of the church. Pete and Charlie watched, but didn't say a word.

After ten minutes, Pete spoke up, "Let's go back to the grave yard behind the church and wait there."

"Why don't you guys go ahead, I'll wait here," said Charlie.

Pete headed around the side of the church to where Davy was snooping around, and together they went over to the cemetery. Billy walked right up to Charlie and put his face right into Charlie's and said, "What is up with you? Why are letting those two geeks hang out with us?"

Charlie spoke without moving, "I like those two, and if you want to hang out with us, then stop calling them names, stop complaining, and stop acting like a bully."

The words seem to hit Billy right in the face. He blinked two or three times and then gave a huge huff. "This isn't right, this just isn't fun anymore," said Billy.

"Then leave, just turn around and leave. I don't need you to be my friend if you feel that way," said Charlie.

Billy was struck with such confusion that it was all over his face. If he left, what else would he do? All the bigger kids wanted to do was stand around and look cool. That was okay, but he really wanted to hang out with Charlie and look through the cemetery. He really liked hanging out with Charlie, and the thought of leaving bothered him. "What is up with you? We hate those guys!" Billy spit the words out of his mouth.

"I don't, I think they're okay. If you want to stay, lay off," said Charlie.

The sound of leaves underfoot made both of them turn around quickly. Amanda had come up on them and they didn't hear her until she was right next to them. They were standing less than a foot apart when she asked, "Is anything wrong?"

Charlie looked stunned, and then realized that his face was very close to Billy's face. He quickly moved away and said, "No, we're just waiting for you. Come on, the rest of the gang is already in the cemetery." He waited for Amanda to step in front of him, and then he turned and said in his nicest voice, "Amanda, this is Billy. He is a good friend of mine. Billy, have you met Amanda? She's new at school this year."

Billy nodded in her direction, and said "hi" under his breath.

Amanda, taking her cue from Billy mumbled, "Nice to meet you." She turned and looked at Charlie, "Let's go to the cemetery. I've never snooped around in an old cemetery before."

They all followed the small path around the back to the cemetery, and they saw Pete and Davy bent over a fallen stone.

"Hey, I think we found something," yelled Pete.

"What is it?" said Amanda.

"Come here and look," exclaimed Davy excitedly, and he stepped back to let the others see.

They could barely make out a round metal disk with a little magnifying glass in the middle. It was full of mud and rust.

"Where did you find it?" asked Charlie as he pulled on it.

"Right here, it's stuck in the ground. We've been trying, but we can't pull it out," said Pete.

At the same time, they all bent down to get a better look. Five faces made a circle around the metal object. Davy stuck his finger in the dirt nearby and started to scrape it with his finger nail. Pete did the same thing. Amanda watched, and then couldn't resist. She, too, started to scrape the mud away. Billy and Charlie followed in, and as they did, the object soon became free. Billy took his large fingers and pulled it up from the ground. He stood up and held it in his hand. "It's just a magnifying glass," he said, and turned it over in his hand.

"Let me see," said Davy and reached for the object.

Billy pulled it away immediately, "get lost," he said and pushed Davy away.

Pete had enough, and stepped in front of Davy facing Billy. "If you ever push him again, I will pummel you!" he yelled at Billy.

Billy stepped forward still holding the muddy treasure, "You think so?" he shot back to Pete.

Amanda looked around frightened. Charlie stepped between the two of them and said, "Knock it off, both of you."

"He started it," said Billy.

Pete glared at Billy, but didn't say a word. He caught the scared look Amanda had on her face, and knew she wouldn't hang out with them if they were constantly fighting.

Davy, too, stepped back and was now snooping in another part of the cemetery. He stopped and looked at the old gate. It had a large sun-shaped sculpture at the very top. It had a face in the center with fiery rays going out from the central circle. One of the eyes had a small crystal, and the other just a round piece of metal. The mouth was smiling, with big oversized lips. It was actually very pretty, but would probably be much more spectacular if it had crystals in both eyes. Pete saw Davy just standing and looking at the gate, "What are you looking at?" asked Pete.

"The gate, look at it for a minute. It's really kind of cool," replied Davy.

Pete looked up at the gate, and nodded his head in agreement, "Yeah, this whole place is great."

Amanda rubbed her hands on the old headstones and read them out loud. "Do you recognize any of these names?" she said to the group of boys. They shook their heads no, and kept snooping, each one in a different part of the cemetery.

She read two or three more, and then Charlie said, "I know that person. I think he's related to my grandmother." Amanda read it again. "Wilber Fairchild, 1835-1887, Father, husband and friend to many." Charlie stood and listened. Amanda moved on to the next stone and read, "Anita Fairchild, 1860-1861, Lovely daughter, be with God." Amanda stopped and stared at the stone. She looked at the place beneath her feet and said, "She was only a baby when she died."

Pete and Charlie stopped what they were doing to look at Amanda. She looked around at them, and then dropped her gaze back to the stone in front of her. She kept reading the stones, but only to herself. She walked slowly to the center of the yard and in the middle was a statue of an angel. It was faded gray and small. It had a ring of nicely arranged stones bordering what once looked like a flower bed. The angel was a little girl with her arms out and her face looking up to heaven. It looked like she was once holding something in her right hand, but it was empty. One wing was chipped, but the rest was in very nice condition. She was standing on her tip-toes, and looked to be dancing. Amanda thought it was enchanting, and stood there looking at it. Davy came up along side of her and said, "I like that statue, it looks so happy."

Amanda turned quickly and saw the peaceful look on his face. She didn't know Davy well, but she liked him from the moment she met him. He was always so agreeable, and genuine. She looked at him and smiled, "I like this place. I thought it was going to be creepy, but it's not," she said.

"Yeah, I did too," said Davy.

They stood there awhile together looking at the statue. Then Davy cocked his head to one side and said, "Does that look like a path there?" and he pointed past the little statue.

Amanda peered in that direction and said, "It sure does. Want to see where it goes?"

"Yeah, let's go."

Davy went first, and Amanda followed behind him. It led off into the woods, then around to the front of the church. "It leads back to the church," said Davy. "Hey, you want to go inside and look around?"

Apprehensively, Amanda questioned him, "You think its okay to go inside?"

"I don't see why not," Davy replied. He walked over to the open window and looked in. "Here, let me make a step," and he put his hands together and bent over so she could put her foot into his hands. She looked at him with wide eyes and said, "You want me to go first?"

"I can climb through the window by myself, but can you get in by yourself?" Amanda stepped back and looked at the others in the cemetery. "What if we all go in together?" she said.

Davy called to the boys, "Pete, we're going to go inside, you want to go with us?"

Pete looked to see where they were, and looked over at Charlie. "Yeah, I'm coming." He and Charlie started walking toward the open window. Billy stood there a moment, and then followed.

Pete was the first one inside, followed by Charlie, Amanda, and then Davy. Billy stood outside and listened to the others walk around inside. He had been in there before, and didn't want to go in. He was still angry at Charlie for inviting the kids he hated most. Billy always felt left out, and he was feeling sorry for himself as he stood outside. He had a terrible feeling that Charlie liked this new group of friends more than him. What was going to happen to him he wondered?

Davy walked around the perimeter where the air was cool and stale. Colored light streamed in from the stained glass windows. The small church was made from limestone blocks cemented together. It smelled like a cave. A statue of Mary holding Jesus stood near the front. There was a cross with cob webs hanging from it, and over the alter was a triangle stained-glass window with a huge, green eye looking in. There were only ten or so pews on each side with shelves for hymnals, but they were all empty. A large book lay open on a pedestal, and a pen was on the top.

They all walked the perimeter of the church, and then Davy stepped back and scratched his head. He was looking at the closed doors. They were stained dark brown, but it looked like small, round indentations were all over them. The indentation made the shape of a bowling pin, or a figure-8. The top was smaller than the bottom, and there were a few larger indentations, four to be exact. They looked like large holes with sunrays drawn around the circles. "Hey guys, look at this," he said.

"Why would someone make a bowling pin on the back of the church doors?"

They all turned to look at the doors. "I've been in here before, but I never remember seeing that," said Charlie. They stood there for a moment, and then they heard a loud thud. Everyone turned to the direction of the sound, and they found Billy lying on the floor near the open window. "What is it?" he said. "What did you find on the doors?"

Amanda started to giggle at the sight of the boy on the ground. Billy looked up at her and glared. "What are you laughing at?" he said in a very mean voice.

Amanda put her hand over her mouth, stood up straight and said, "I'm not laughing at you, I'm laughing with you."

Billy got up quickly and walked over to her. He stepped right inside her 'space' and was going to stare her down, when it happened. She stood her ground and looked right back into his eyes. He was defenseless, everyone was looking at him. He stood there for a moment, motionless, his shoulder stiff, and his eyes fixed. He was larger than her, and was intimidating just by size alone. He waited for her to cower, but instead she stared back into his red and angry face. Billy realized that the other boys were still watching him, and he must have looked like a bully. He stepped back, and his whole body relaxed, and he said something that only Amanda could hear. "Sorry to be so rude. I'm just a little jumpy today."

Amanda's smile became soft and she replied, "I didn't mean to offend you, I'm sorry too." From that moment on Billy Martin became a much nicer person. He filled his lungs with the stale, damp air, and moved to look closely at the back of the church doors. "I've been here hundreds of times, and I've never seen that," he said.

"That's because you never were with me," said Davy.

Billy looked over at Davy with an evil glare, but it faded as he looked past Amanda. Davy didn't want to push his luck, so he backed up and gave Billy the right of way.

"Why are you so interested in what is on the back of the door?" said Pete.

Billy turned and glared at Pete. He still did not like Pete at all. He wasn't going to divulge any information to him, not now, or ever. "None of your business," said Billy back.

Pete raised his hands and walked away.

The group spent another half hour looking at the church. Billy examined the door, putting his fingers in the little holes and tracing the figure-8. Davy admired the ornate hinges on the door. Amanda looked at the beautiful stained-glass windows and marveled at how the light lit the inside of the church. Pete and Charlie looked at the alter, and the book that lay open. It was getting to be lunch time and Charlie looked up and said, "Hey, what do you guys say we go back to my house and have something to eat?"

They all nodded in agreement and headed for the open window. One-by-one, they maneuvered through the window, and walked the path back to Charlie's yard. Near the end of the path, Charlie caught up with Billy and said in a low voice, "What's the mystery with the old church?"

Billy stopped and turned around, "You don't know? You live in that house and you don't know?"

"Don't know what?" said Charlie.

"The fortune, the lost fortune is said to be somewhere near that church," said Billy.

"What fortune? What are you talking about?" demanded Charlie.

"Ask your mom, your stepdad, or even your grandmother. Everyone knows," and then he came close, so that his voice was not overheard, "and the secret is in that church."

Billy turned and headed down the path to Charlie's yard. His hand felt the hard, round object in his pocket. It was the magnifying glass they found in the yard. He had a feeling that this was something that would lead him to the fortune.

Pete, who was standing on the path just a few feet away, heard Billy tell Charlie of the fortune. He heard them coming and turned quickly to catch up with Davy. A fortune is in the church somewhere. He couldn't wait to get back and tell Davy.

CHAPTER FOURTEEN

ALL THAT GLITTERS

The thought of a fortune brewed in Pete's mind. Billy had been so insistent that he couldn't get it out of his head. What if Billy really knew something? What if he was right and there really was a fortune to be had. The very thought lingered in and out of his thoughts for the next two days. It finally surfaced one afternoon, when they were sitting by the pond eating ice cream. Pete said off the topic, "What would you do if you had a fortune?"

Davy piped up, "I'd buy my own island, my own country, no my own continent and be the President. No one would ever tell me what to do. I would have a big army, and protect my borders, and if anybody tried to tell me what to do, I would blow them off the map."

"Very nice, very nice," interrupted Pete, and patted his friend on the back. "Amanda, what would you do if you had a fortune?"

Amanda was taken by surprise. "I don't know, I never really thought about it."

"All girls say that," said Charlie. He raised his voice an octave and said, "All I want is World Peace, Love, and Understanding," and as he said it, he moved his arms up and down.

Amanda glared at him, "That's not what I meant," she said.

"What did you mean?" said Charlie, and he leaned in close swooning over her.

"I meant," and she said it with emphasis, "I don't know if I would live in a big house or a mansion," and she gave a little giggle. "You know, a house like Pete and Charlie live in, or a house like the Queen of England lives in." She sat up and extended her pinky, and made a sipping motion.

Pete smiled at her, and said, "I would like to travel, the whole world and see everything."

Davy replied without thinking, "like your dad?"

Pete's smile immediately vanished. Davy looked at him and said, "I'm sorry, I didn't mean...."

"It's okay, you didn't say anything wrong," said Pete back to Davy.

Charlie jumped in to ease the tension, "I would build a restaurant. Some place nice, where you could get a really fancy meal. This town doesn't have a decent restaurant. I bet I could make an even bigger fortune with a good restaurant."

Behind the bushes, where he couldn't be seen, Billy sat and listened to the others in the group. He didn't like that Pete knew about the fortune. He was wondering if Charlie had told him. His anger was beginning to boil.

"Why are we talking about fortunes," asked Amanda.

Charlie suddenly realized that he hadn't told anyone what Billy had told him. How did Pete know about a fortune?

Pete, without realizing it, had brought up the subject. He had to come up with something, or he would have to tell how he knew. He took a deep breath, but before he could say a word, Davy was talking. "The people in this town are always talking about the fortune in the old church. The same church we were all in the other day." He began to ramble on about how people were always discussing the location of the fortune.

Pete interrupted Davy, "What if the fortune isn't money? What if the fortune is an old painting or a land trust that's already been sold?"

Suddenly, Billy rolled forward from his hiding spot in the bushes and burst out, "A fortune has to be money. What else would it be?"

Charlie and Amanda jumped up in surprise. Davy fell backwards and Pete let out a yell. They froze shaken by the surprise that Billy was hiding in the bushes all along.

"I, I, I agree with Billy, a fortune is always money," said Davy quietly in fear of the larger boy.

Charlie, still shocked at Billy's behavior, rubbed his chin and looked at the rest of the kids, "You know, Pete's right. A fortune is different to different people. It could be called a fortune, but really be worthless."

"I don't care. I'm still going to try to find it," said Billy.

"I, I, I want to find it too," chimed in Davy, and he looked squarely at Billy. It suddenly dawned on him, that for the first time in his whole life, he agreed with Billy. Maybe it was because they lived here in this small town for so long, and they were convinced there really was a fortune.

Amanda, who had been silent and listening to the group, stood up and wiped her hand on the back of her shorts. "I'm in," was all she said.

Billy gave a grunt, and looked at Charlie.

"I'm in," said Pete, and looked from Amanda to Davy.

Billy snorted and said, "How do I know I can trust you? How do I know I can trust any of you?"

Pete found this amusing coming from Billy. He let out a small chuckle, and that was enough to get Billy off the ground and to his feet. He was in Pete's face faster than you can say one, two, three. Pete did not see it coming, and immediately jumped back.

"Knock it off," yelled Pete.

Charlie quickly put himself between the two boys and stared Billy down. "We are going to have to trust each other if we are going to work together." He looked around at all the faces, and then back to Billy. Billy's face was red, "How do I know you guys aren't going to leave me out? How do I know you won't go out to the church without me?" Billy's insecurities were showing, and he felt vulnerable. Charlie got closer to Billy and said in almost a whisper. "Hey buddy, I haven't left you out yet. It really is up to you."

Billy, feeling put on the spot, took a step backwards and looked into the faces in the small group. Feeling apprehensive, yet afraid to be left out, he shook his head yes and said, "Okay, we work together."

The tension of the small group was still thick. Amanda, feeling suddenly tired said, "Hey, I think I am going to be heading home. I will see you all at school tomorrow." She turned to leave and Pete jumped up. "You mind if I walk with you?" He turned to look at Davy, and he too leapt to his feet.

The small group abruptly broke up, and three of the five were walking quickly away. Charlie and Billy were left standing by the pond. Charlie, disappointed to see everyone leave sighed, "Billy, you really do know how to break up a party."

Billy, not sure what just happened, walked with Charlie back to his house. He still wasn't convinced that he was a part of this small group. Nor was he sure that he wanted to be, given his mistrust of Pete and his bad standing with Davy.

CHAPTER FIFTEEN
THE ODD FIGURE-8

It was strange how the figure-8 on the back of the doors at the old church was similar to the diagram Pete found in their science textbook. While Pete and was waiting for the rest of the class to finish the math, he was thumbing through his science textbook, and he saw it. "Hey, that funny figure-8, it's here!"

Charlie looked up from his math assignment and said, "What 8 is here?"

"The figure-8 that was on the doors in the back of the church. A picture of it over a house is in this book, see." He turned the book so that Charlie could see the picture.

Charlie peered at the picture and his expression changed. Amanda, who just looked up saw Charlie's face change and said quietly, "Let me see." Pete moved the book her way.

The picture was of a house with a bright figure-8 above it. It looked like the sun made a path with large bright spots in the form of a bowling pin in the sky. "Isn't this the same shape that was on the back of the doors in that church?" asked Pete.

"Yeah it is," said Charlie. "I wonder what this shape means."

Pete put his head closer to the caption to read out loud, "The shape of the analemma is created by the noon sun throughout the year." It didn't give him enough information, and so he read the whole page quietly to himself. Not satisfied with the explanation, he said, "I think I am going to take the textbook home and ask my grandma what she knows about this."

Charlie was just about to comment, when the teacher told the class to put their math away, and take out their science textbooks. The lesson in science began, but Charlie, Pete and Amanda were on a different page. The picture on the page was exactly like the indentations on the back of the doors in the church. Pete's adrenaline was flowing. He couldn't wait to tell Davy.

At the end of the day, Charlie packed his science book in his backpack and said, "I think we should all take our science books home. Pete, ask your grandma, I'll ask my mom, and Amanda, you look it up on the internet." They all nodded in agreement, and left for the day.

Pete couldn't wait to get home. He wanted to show his grandma the picture and he was sure she could tell him the secret of the odd-shaped 8 diagram. When Pete walked in the house, it was dark and quiet.

Pete's grandma walked out of the shadows of the hallway. "Shhh, your Grandfather isn't feeling well, he's lying down to take a nap," she said. "What is it that you need Pete?"

"Hey Grandma, we found this picture, and this is the second time we saw this," he handed her the textbook opened to the page, "We were wondering if you know what this is?" said Pete.

She looked long at the picture, "this," she said, "is the noon sun throughout the year. Everyday the sun is actually in a different place. We can't notice it with our eyes, but given a backdrop like this one, and photographed every other day or so, then this is the shape it takes. This is called an analemma. They used to put them on globes."

She left the room, and Pete could hear her in the study moving things around. When she came back in, she had a globe in her hand. It was old and yellow looking. She sat down at the head of the table and motioned for him to follow. "See this," she pointed to a shape similar on the globe as in the picture. "This analemma has numbers and months written on it. It shows where the sun is located at any given time of the year." She adjusted her glasses to look closely at the globe. "Let's see," and she got very close to the surface of the globe. She pointed her slender finger to the top and said, "Right here, today, the sun is located at 15 degrees south latitude."

"What does that mean?" said Pete.

"It means that the sun is below the equator at the 15th parallel. It will remain below the equator, getting closer to the Tropic of Capricorn every day. When it reaches the 23rd parallel, then we will have the winter solstice, the shortest day of the year for those of us in the northern hemisphere. It will mark the beginning of winter. Our days will be much shorter than our nights."

"So that's why our days and nights change length," said Pete. "This tells us where the sun is located at noon, and above the equator our days are longer than nights, and below the equator, our nights are longer than days."

"That's right," she said.

"Then Grandma, why isn't it just a straight line up and down? Why does it make that 8-shape?"

"Well, you see, the Earth does not travel at a constant speed. Sometimes it speeds up, and sometimes it slows down. Didn't you do a presentation on Kepler?" she asked him.

He shook his head up and down. "What does Kepler have to do with this?" wondered Pete.

"Kepler's laws of planetary motion, let's see, which one is it that says that a planet covers equal area in the same time."

"That's Kepler's second law," said Pete.

"That law says that if the planet is going to cover equal area, it must speed up as it sweeps close and slow down as it moves farther away. It is this change in speed that causes the 8-shape," said Grandma.

"So it makes a funny shape, it can't mean much," said Pete a bit disappointed.

"Actually it does," said Grandma. "If you go outside at noon, on most days, the sun is not at your celestial meridian. When the sun crosses your celestial meridian, then it is called high noon. When the sun is on the right side when your watch says noon, this is called slow sun, or sun behind. This means that solar noon could be 1 to 16 minutes later than clock noon. When the sun is on the left, it is called fast sun, and solar noon may occur 1 to 16 minutes before clock noon. Only four days of the year do solar noon and clock noon occur at the same time."

Pete looked at her trying to take it all in. "So when my watch says noon, it's not really noon according to the sun, on most days. Is that right?" he asked.

"That is right," she responded. "One of the four days that solar noon and clock noon agree is December 25th."

"What a coincidence," said Pete.

"You really think so?" said Grandma, and she smiled and winked. "Most historians will tell you that Jesus was born in the summer, and the date of December 25 was chosen to celebrate his birth."

Pete's eyes grew wide, "I bet they picked that day because the sun was at the celestial meridian."

"I think you're probably right," said Grandma.

Pete dropped his hands into his lap. "How am I going to explain this to Charlie and Davy? I barely understand it myself," said Pete.

He took the globe to school with him the next day. When he explained it to the group, Charlie and Amanda caught on quickly and nodded with understanding.

"I don't get it," complained Billy. "I think you guys made this up." His insecurities were coming through.

Davy, too, shook his head. He didn't say anything, but he wouldn't make eye contact with Pete.

"Charlie, did you find out anything from your mom?" asked Amanda.

"My mom never heard of such a thing, and neither did my stepdad, but he was interested in the shape."

"Amanda, did you check the internet?" asked Charlie.

She shook her head no. "I couldn't last night, but I will try tonight."

Charlie looked at the members in his group. "I think this could be a clue, we really should find out more."

Amanda and Pete nodded in agreement, Billy scowled and Davy looked out the window.

CHAPTER SIXTEEN

TRAGEDY FALLS

Always without warning, and always without regard to the people involved, tragedy happens.

Monday was a day like every other. Charlie decided to ask the teacher if she knew anything about the analemma. He was going to exhaust all resources before giving up. Pete was watching Amanda while Charlie was talking to the teacher. Amanda was holding his grandma's sun clock and trying to get the arm to come up. They had planned on using the sun clock to help demonstrate the analemma. She pushed in the compass, but the arm would not pop out. Finally, she held it up and shook it violently. Pete's eyes went wide and he said, "What are you doing?" and reached for it.

She put out her other arm, to block him from grabbing it out of her hands. She found something. She put the sun clock to her ear and shook it again. It made a rattling noise. "Hear that," she said, "there's something in there." She handed it to Pete, and he shook it.

"What's in there?" she asked.

Pete raised his shoulders and eye brows, "I dunno, I can't figure out how to get the back panel off." He pushed on the clear globe and the arm popped up. "How did you do that?" she said.

He pulled the string and attached it to the end. The sun clock was ready to go. Amanda turned away to face Charlie, and Pete held the clock up to his ear and shook it again. He was sure there was something inside, but could not figure out how to get to it. Maybe he could ask his grandmother, since it was hers.

"Hey, you know what we should do?" Pete said. "We should go back to the church and get a look at the door. Maybe we could figure out what it means."

Charlie's eyebrows went up, "I think your right. Want to go after school today? Can you guys make it?"

"I'm in," said Pete.

"Me too," said Amanda.

"Let me see what Billy says," said Charlie.

"I'll ask Davy at lunch," said Pete.

By three o'clock, it was a go. Everyone was going to go home and get a flashlight, and then meet back at the church as soon as possible.

Pete got off the bus and rushed his sisters to the house. They were in no mood to be rushed and resisted the fast pace home. By the time Pete got into the house, he was in a fowl mood and wanted to quickly get the flashlight and go back out again.

Pete found the flashlight, but when he was at the back door, he heard his name being called from upstairs. "I'll be back in about and hour," he yelled up the steps.

His grandfather appeared around the corner and was saying his name. While he was talking he took his cane and rested it on the top step. "Pete, your mother and grandmother went to pick up some groceries, and asked if you would hang around just a few minutes until they got back home," he said.

Why today of all days? "Grandpa, do I really have to? I made plans to meet my friends at the old church, and if I don't go…" he trailed off.

Pete's grandpa was waving his free hand about his ears, "You can go as soon as they get back, now stop the whining," he said.

Pete huffed and stomped his foot in anger. Of all the times to have to baby-sit his sisters, why now? He was really mad, and he slammed the flashlight down hard on the kitchen table.

"Pete!" yelled his grandfather, "Now you settle down," he said, and reached for the hand rail to start down the steps. His hand missed the rail, but his foot started down anyway, and without the brace from the rail, his grandfather lost his footing and started to tumble head first down the steps.

Pete heard the sound of his grandfather falling. For a split second he thought it sounded like a large bag of potatoes rolling down the steps. He secretly hoped it was potatoes, but he knew it wasn't. He moved to the foot of the steps, and saw his grandfather hit the last step, and then lay motionless at the bottom.

Pete held his breath and then screamed "POPS!" He moved to him quickly and cradled him in his arms. He touched his forehead, his cheeks, and he rocked the old man in his arms back and forth. Suddenly it occurred to him that he needed to call for help. He gently laid him down and grabbed the phone on the wall. He dialed 911, and calmly spoke to the operator. When he hung up, he went back over the body lying still on the floor, and the gravity of the situation hit him. His face got hot, and his eyes welled with tears. He bent down and again cradled him in his arms. "Grandpa, grandpa," cried Pete, with tears flowing out of his eyes. He gently touched his forehead and swept the gray hair back over his head. His tears dropped on his grandfather's face as he rocked him gently in his arms.

The two little girls stood in the door way and cried softly holding each other as they watched their older brother hold their grandfather. It seemed like hours, but it was only minutes before the house was swarming with people. First the paramedics arrived, then his mother and grandmother, and then the neighbors. Faces of people he knew and didn't know. Questions and more questions were asked. Pete was in agony knowing it was his fault that his grandfather fell. He was sure everyone could see it in his face, the guilt and shame of his bad behavior.

Pete was sitting at the kitchen table staring off into space. A large glass of water sat in front of him, and he could see the sky turn pink as the sun set on the horizon. His grandmother and mother went to the hospital, and the neighbor was watching his sisters. He heard other voices in the house, but he didn't know who they were. He repeated the fall over and over in his head. It must have been his fault; it had to be his fault. He was the one having a temper tantrum. His grandfather probably wouldn't have come down if he wouldn't have acted the way he did. What was he going to do without his grandfather, and again large tears flowed down his cheeks. How could this have happened?

The door slammed, and Pete jumped. Voices, but this time he could tell who they belonged to. It was Davy and Charlie. They took one look at Pete and slid into the chairs at the table. No one spoke for a long time. Then Pete took the back of his hand and wiped off his cheeks. "It's my Grandpa, he fell down the steps. I think he might be dead."

Davy shook his head and Charlie spoke, "Do you want me to call the hospital?"

Pete focused his blurry eyes on Charlie's face, "could you?" he asked.

"Sure," and he walked over the phone and dialed. He spoke in a hushed voice, and then hung up. "They admitted him, he's critical, but he's hanging in there," he said.

Pete looked up at Charlie and his eyes said 'thank you' and Charlie nodded.

He didn't remember changing his cloths or even going to bed. When he woke up his first thought was about what he was going to do that day in school. He had forgotten about the tragic event the day before. It wasn't until he went down to breakfast to find five strange people sitting around the kitchen table. They all looked up at him when he came into the room. His mother, his grandmother, and three other people he only vaguely knew. They were all red-eyed, and it hit Pete at that moment that his grandfather must have died. He didn't say a word, but it was confirmed by a nod of grandma's head when he looked over at her. A huge lump swelled in his throat, and his mother gathered him in her arms. Again, his eyes leaked large drops, and he felt them hit his arms as they rolled off his cheeks. He wanted to run back up to his room and hide under the covers.

His mother led him over to the table where he sat down and began to listen to the conversation that was already in progress. He was vaguely aware of the coming and going of the neighbors. He didn't go to school that day. In the quiet moments, when the house was still and no one was around, Pete became overwhelmed with guilt. It occurred to him that he could not always be the center of the universe, and that in his selfish moment, he had caused a tragedy.

That evening, when the house was still and the neighbors had left, Pete's grandmother came into his room. She sat down gently on his bed and took his hand in hers. "This is not your fault," she said.

Pete looked up at her pale green eyes and said, "Grandma, I was being loud and mean, and he was coming down the steps to talk to me. How can this not be my fault?"

Grandma took Pete's hand and gently put it up to her face. This dear child was overcome with grief, and to ease his pain she said, "Pete, your Grandfather was very ill. Even though he didn't show it, he was suffering. This is no one's fault; it's just hard to lose him." She gave a small smile, and closed her eyes. "He was my world, and I adored him. I am going to miss him each and every day," she said, and a small tear rolled slowly down her smooth cheek. She looked like an angel at that moment, and Pete understood the meaning of real love for the first time in his life.

"Grandma, I am so, so sorry," cried Pete, and he buried his head in her shoulder.

"No need to be sorry, it was meant to be," she said. She pulled him up and looked him in the eye. "Now my dear, you are the man of the house. Until your father returns, and I hope he comes home soon, but until then, we are going to need your help."

Pete nodded at her words, but they did not sink in, not yet. She stayed a few more minutes, then hugged him tenderly and got up. "I want you to wear your best suit tomorrow, you are going to help carry him," she said. She rubbed his head, and then left the room. It was so lonely when she left, he wished she could have stayed and held him longer.

Pete slept fitfully that night. When his body finally went limp, he drifted off into a painful dream. *It was dark in the house, very little light to see. He heard a soft noise downstairs, and headed down. When he reached the bottom, a small yellow light was coming from the living room. He walked in and saw the curtains in the room billowing up and touching the ceiling. It made this soft brushing sound. No one was in there, but there was a large wooden box sitting on the floor. He walked over to the box and opened it. All that was there was a flashlight, granddad's old cloths, and a pair of boots. He closed the box and turned to leave the room. It was suddenly full of people wearing black and crying.* Pete sat up in bed, sweating. He looked at the clock and it was 7:30 in the morning. He got up, showered and got dressed.

Pete was uncomfortable in his suit, and wished he could just skip the whole affair. He wanted to stay home in his room where it was safe.

The ride to the funeral parlor was too short, and ended sooner than he expected. He walked inside the chapel and went straight to the casket. He looked into the face of the man he came to know and love so much. How could he be gone?

It took every ounce of strength to stand straight and greet neighbors and friends. He nodded, smiled and made small talk. When it came time to carry the coffin, he was in the front, and the weight of the coffin was as heavy as the grief he carried inside of him.

The ride to the grave yard was also too short. He stood next to his mother and she held the hands of his little sisters. His grandmother slipped her hand into his and gave it a squeeze. The words of the preacher seemed to dance around his head, and it suddenly hit him, he was the only male standing in this small group. He took a deep breath and let it out very slowly. Today, and from now on, he would act differently. He knew he would stand taller, say 'yes' instead of yeah, and help without being asked.

A black sedan with tinted windows moved slowly up the road during the final part of the burial. It stopped briefly, and then slowly moved on. Pete was getting the feeling that someone was watching them.

CHAPTER SEVENTEEN

MORGANITE

Life is suppose to test you, change you, make you better. Today, Pete was determined to be better. Having his granddad in his thoughts, he made it his mission to help get the girls ready for school. He took charge of loading their lunch boxes and setting off on time. He wanted to show his mother and grandmother that he was going to start being in charge.

The day was harder than he expected. The teacher was especially nice, and so were the other students. Pete suspected that the story of what happened that day after school must have circulated, and he was embarrassed. At lunch, the conversation started slowly. Everyone was unsure of what to say to Pete. Aware of the awkward moment, Pete asked what they had found at the church.

Before they found out about the accident, the four other kids had entered the church through the window and went to the door to inspect the analemma in the back of the church. They found not only one large hole, but four. They also found that the larger holes were odd shaped, like octagons, while the other holes were just circular. Davy suggested that it could be the same stone that was in the eye of the sun on the gate to the cemetery. They looked at it, but without taking it out, they couldn't be sure. Billy wanted to take it out, but Amanda stopped him. She said it would be vandalism, and Charlie agreed. They decided to go back to the church and grave yard with flashlights and search for stones that could fit into the larger holes in the analemma.

Pete, still very sad, decided that he was up for the adventure, but said he had to make sure he wasn't needed at home. The other kids nodded their heads with understanding. Pete was almost ready to leave, when he remembered that he had his sun clock and sexton in his locker. He wanted to take them home, and return them to his grandmother. He wasn't going to make off with other peoples' things any more. When he reached for the sun clock, it felt heavy in his hands, and he heard the faint rattle of something inside. When he gave it back, he decided to ask his grandmother if she could open it and show him what was inside.

Amanda sat next to him on the bus, just like she did everyday. Today's ride home was solemn. Pete wanted to talk to her, but all of his words were lost in his head. He couldn't bring himself to speak. Amanda, being very intuitive, sat next to him in silence. Sometimes it just felt good to be quiet with a friend.

Before she got off the bus she said, "Pete, I am very sorry about your grandfather. I know you were close to him." Pete could only nod his head for no words could form. "Pete, will I see you at the church?"

Pete looked at her with very little expression. He really wanted to join the gang, but he needed to be around if someone needed his help. "I dunno, maybe," he said back.

Pete walked the girls' home, and they were silent too. Pete wondered how long this passive quiet would last. When they entered the house, Ashlyn caught the back door so it would not slam shut. Pete's grandmother was standing at the sink staring out into the yard. When they entered, she turned slowly to face them. "Did you have a nice day at school?" she asked.

The two little girls responded a quiet "Yep." She smiled and put her arms out to them. Pete waited for the girls, and then he also hugged his grandmother. A look of longing was all over her face. "Grandma, you Okay?" She hugged him again and said, "I've had better days," and gave a weak smile. Pete put down his book bag and poured himself a glass of water. He walked over to the table and started emptying the contents of the bag. He pulled out the sexton and the sun clock and put them both on the table. His grandmother watched him and said, "I'm glad you take such care with those."

Pete looked up at her, "here, I want to give them back to you." He handed both of them to her, and her expression changed. "No Pete, I really want you to have both of them. I know you like them as much as I do, and I would be happy if you would hang on to them." She pushed them back at him. It was then that Pete felt the shift inside the sun clock. "Grandma, does this panel on the back come off?"

She took the sun clock in her hands and started to push the panel in different directions. "I think it does, I really can't remember," she said.

The phone rang, and Pete went over to answer it, still watching his grandmother fiddle with the sun clock. It was Charlie and he was asking if he was going to come. Pete said he would ask if they needed him, and if not, he would be right over. He hung up the phone and turned around, and when he did, he saw the back of the clock was open. She handed it to him and said, "I do remember how it opens."

Pete looked inside, but it was empty. "Grandma, wasn't there something in there? I mean, I swear I could hear something rattling around when you shook it."

She had a very strange look on her face, and her left hand was clenched into a fist, "No, there doesn't seem to be anything inside," she said. She slid the panel back on and handed it to him.

He watched and held out his hand to take the sun clock. "Can I go hang out with my friends?" he asked.

"I don't see why not, unless you have homework to do," said his grandmother.

"I got all of my homework done at school," said Pete, and his gaze went to her left hand. He was sure that whatever 'it' was in the back of the clock was now clenched firmly in her fist. "I'm going to put my stuff upstairs and change shoes, and then I'll go," he said. He left the room, but stopped in the shadow of the hallway. From there he could still see his grandmother.

She waited a few minutes, and then went back over to the window. She put her fist up to her face and opened her hand flat. The small crystal lay sparkling in her hand. She took her other hand and rolled it around, and then slipped it into the pocket of her apron. Pete watched in amazement. Could it be that the crystal they needed to find was already found. All he had to do was watch where she put it, and then take it and see. How was he going to get it out of her apron? If he left now, then she might put it someplace where he would never find it.

He went upstairs and put his book bag down. He changed his shoes and sat heavily on his bed. He needed to stay here and watch what she did with the crystal. He grabbed his math book and some paper and headed back down to the kitchen. He would tell her that he realized he did have homework and needed to get it done.

She was still in the kitchen when he came back down, and so he slid into one of the chairs, opened his book and started working problems. "I thought you were heading off with your friends?" she said to him.

"I looked at my assignment sheet, and I realized that I missed something. The responsible thing is for me to do my homework first, and then go hang out," he said to her.

She came over to him, hugged him to her chest, and then left the room. Pete needed to follow her and see where she was going. He watched her go down the hall from the seat he was in, but she disappeared into one of the rooms off to the left. Pete slid quietly out of the chair, and tip-toed down the hall. She was sitting in her favorite chair in the living room working on needle point. The stone was no where in sight. He did see a slight bulge in the pocket of her apron, and figured it must still be in there. He very quietly returned to his spot in the kitchen and continued to work on his math. His mind raced back to what the others had said today at lunch. What if this was the very stone they were all looking for now?

Pete kept a very close eye on his grandmother all through out the evening. He helped her make dinner, bath his little sisters, and clean up the kitchen. His mother had a job that kept her away a few evenings a week, so it was just him and his grandmother now.

When the twins were tucked in bed, Pete sat outside with his grandmother on the porch. The air was starting to get cool, and a breeze swept over the ground. Leaves shifted, and fell off of the trees. It looked like it was raining leaves. His grandmother was rocking slowly with her eyes closed. She was humming quietly to herself. Pete looked at her and said, "Grandma, have you ever been to the old church over by the little cemetery?"

Her eyes opened, but her head was still back on the chair. Her rocking slowed down and she said, "Pete have you been to that church?"

"Grandma, I asked you first."

"Yes I have. Now have you been there?"

"Of course I've been there, that's why I'm asking," he said.

"Is there something about that church you want to ask me?" She came to a stop and lifted her head off the back of the chair.

"What do you know about that church? I mean, do you know about the fortune that's suppose to be there?" he asked.

This question made her chair rock forward and stop. "What did you say about a fortune?"

Pete froze, he gazed directly into her steady eyes, "Uh, well, you see, one of the kids I hang out with said that there was a fortune and the key was in the church."

"Oh," she said, and started rocking again. She closed her eyes and put her head back.

"Grandma," said Pete, "you know something and I know it."

She spoke without opening her eyes or lifting her head. "I know something about that church, and I know something about a fortune, but it doesn't matter now."

"What do you mean, it doesn't matter now?"

"I believe the fortune is gone, lost, never found, and believe me, many people have looked for it," she said. "I heard about it when I was young, and no one ever found anything."

"I think we found something," said Pete.

Again, she stopped and leaned forward, "Oh yeah? What have you found?" she said.

"We found an analemma on the back door of the church."

She looked at him in surprise, "you found the analemma?" she repeated.

Pete suddenly realized that she knew more than she was willing to tell. "Yeah, and we think we know what we need next."

"What do you need next?" she asked.

"We need a crystal, like the one you have in your pocket," Pete ventured.

Her hand went down to her apron and she could feel the bulk of the crystal still in her pocket. "I saw it when you took it out of the sun clock." She pulled it out of her pocket and slowly opened her hand. Pete looked at it. It was round and perfectly facetted, about the size of a quarter. It was clear, but pink in color.

"Is it a diamond?" asked Pete.

"NO, this is Morganite it comes from the mineral beryl. If a beryl is green, then the stone is called emerald, and if the beryl is blue, then the stone is called aquamarine. They are precious in the gem world," she said.

"Is it worth a lot?" asked Pete excitedly.

"Yes, yes it is. You see why I couldn't let you just have it," she replied.

"Can I hold it?"

She reached over and gently dropped it in his hand. "You think this is the stone you're looking for?"

"Maybe, yes," he replied.

He held it up and looked through it. It was almost completely clear with a hint of pink. It was beautiful to behold. He handed it back and said, "Is it the stone we're looking for?"

"Maybe, yes," she said.

"Can I have it?" said Pete.

She looked at the stone and slowly handed it to Pete. "You must be very careful," she said.

"Oh, I will. I won't lose it, I promise."

"That's not what I meant," she said. Again she leaned forward and her gaze pierced directly into him. "I mean, be careful because there are others who look for the fortune. Other people that could hurt you if you get too close."

The flesh on Pete's arm went to goose bumps. The way she said it made him scared. What was it that she knew that she wasn't saying? "Grandma, you know don't you?" he asked her.

She leaned back, closed her eyes, and rocked slowly. She said nothing for a long moment. Then, without opening her eyes or lifting her head she said, "Isn't it time for you to go to bed?"

Pete knew that the conversation was over, and he wasn't going to get any more tonight. He looked at the stone in his hand and was filled with a new found joy. He had the stone, but what it meant he didn't know. He did know that he would have to figure it out for himself. He got up, and went upstairs.

CHAPTER EIGHTEEN

NUMBERS AND LETTERS

The name of the crystal sounded like a town in Wyoming, but it's a pale pink stone. It's also very valuable, and Pete knew it. He wrapped it in a small piece of tissue paper, and put it back into the small compartment in the sun clock. When he put the sun clock in his backpack, he eased it in with more care than ever before. He couldn't wait to tell the others that he had the stone they may be looking for.

As soon as Charlie and Amanda were seated at the table, Pete bent forward and said, "I have it."

"You have what?" asked Charlie.

"I have the stone you were looking for at the church," said Pete.

"How could you have the stone, you weren't even there yesterday," said Charlie.

"I don't know how, but I have it," said Pete. He took out the sun clock and carefully slide the panel to the secret compartment back, inside was a small piece of tissue paper. He picked it up, removed the paper, and carefully put the stone in the center of his palm. "It was in here all the time," said Pete.

Amanda's eyes were very large. Charlie's mouth opened and his eyebrows raised almost off his head. "How do you know this is the stone we've been looking for?" questioned Amanda.

"My Grandma said it was," said Pete.

"How does she know that?" said Charlie. "Did you ask her?"

"As a matter of fact, I did," said Pete. "She knows more, but she ain't saying. But, she did say this was the stone we needed," said Pete.

"Is it, is it a diamond?" asked Amanda.

"That's what I first thought too," said Pete. "It's called Morganite, and it's valuable. It's the same mineral that emeralds come from."

Amanda's eyes were wide in disbelief. "Can I see it?"

Pete carefully handed it to her. She took it and held it up to the light. "It's really beautiful," she said. "How many carats do you think it is?"

"What's a carat?" asked Pete.

"It's a way they measure how big a stone is," said Charlie. "Can I see it now?" he asked. Amanda handed it to Charlie, and he too held it up to the light. "It looks like the right size, but how does your Grandma know this is the stone we're looking for?" said Charlie.

"I can't tell you how I know, but I just do. Grandma is really knowledgeable about a lot of stuff. I wouldn't be surprised if she hid the fortune, or was there when someone else did," explained Pete.

"Why don't you just ask her if she knows?" asked Charlie.

"I did, but she wouldn't tell me," said Pete. At that moment the teacher began talking, and so not to be seen with the stone, Pete quickly wrapped it up and put it back into the small compartment of the sun clock. He didn't notice the teacher heading over to his table. When he turned to face her, she reached in front of him and picked up the sun clock. Pete watched in horror as she said, "Mrs. Newman's class is studying Kepler, and I told her of the interesting devises you had to measure the sun and keep time. She asked if she could borrow them to show her class. Pete, you don't mind if I lend them to her, do you? She will take very good care of them, and return them soon."

Pete's mouth opened, but no words came out. It was as if he was dumb struck. Charlie saved the moment, "Ms. Nagel, Pete's Grandma asked Pete to bring these back home today. They belonged to Pete's grandfather"

"Oh, of course, I understand," she said, and handed the sun clock back to Pete.

"Thank you," he whispered to her, and put the sun clock carefully into his backpack. "Thank you," he mouthed at Charlie. Charlie gave him a thumb's up, and a wink.

At lunch, they laid out the master plan. Everyone had to come to the church after school. Pete was to bring the sun clock and more importantly the stone.

The day seemed to drag on endlessly, and Ms. Nagel heaped on the homework. On the bus, Pete was groaning about how much they had to complete for tomorrow. Amanda looked at him and said, "Peter, stop whining about school. We have a fortune to find. Do what you have to at home, and then get your butt over to that church. We are all counting on you."

Pete could only stare at her. She was even prettier when she was bossy. "Okay," he said weakly.

She got up to get off the bus and said, "You will be there, right?"

"Sure, yeah," he said back.

Pete ran to the cabinet on the porch and pulled out the big flashlight. He put his book bag in the corner in the kitchen, and pulled out the sun clock. "Grandma, today I really have to run. Be back by dinner, bye," and he was out the door letting it slam behind him. "Sorry," he yelled as he ran down the sidewalk.

She shook her head. "Pete, what am I going to do if you find it?" she said to herself.

They all made it to the church in record time. Billy was leaning next to the window biting his fingernails. He shot Pete a mean look as he passed, but did not utter a sound. One at a time they slipped through the window. Billy was getting more stealth, and didn't make such a loud thump when he hit the floor. They turned on their flashlights and ran to the back of the church. Pete quickly retrieved the pink stone and shined his light on it so everyone could see it. One at a time, they passed it around the circle so everyone could take a look at it.

"It's so big," said Davy.

"I wonder if the fortune is more stones," said Billy.

"I would like that," said Amanda. They all turned to look at her. "What?" she exclaimed as she blushed.

Pete was the last in the circle, and held it up for everyone to see. He placed his hand up to one of the larger notches in the door, and put the stone easily inside. The notch was cut perfectly to house the stone. It gleamed beautifully in the door. They stood there and looked at it, and then Charlie spoke, "so what does it mean?"

They looked at each other with anticipation. "I wonder if it fits in the other notches," said Pete. He carefully removed it and placed into another large notch, again it fit perfectly. He tried the remaining two, and again the stone fit perfectly in all four.

"Are there supposed to be four stones all together?" asked Amanda.

"I don't know," said Charlie. "Why don't you ask your Grandma where the other three stones are?" he said to Pete.

"Okay, we find the other three stones, but what does it mean after we find them?" wondered Pete.

"Well, maybe it will make a wall shift back, or a pew slide over, or an opening in the floor with steps leading down to a secret tunnel," speculated Davy.

Pete looked at him, "Come on. Can you be serious?"

"What if he's right?" said Billy. "I like those ideas."

Amanda put her hands in the air, "Okay, so we're standing in front of the biggest clue, and we don't know what it really means," she said. "We need to lay out what we know, and see if we can come up with more ideas."

Pete sat down in the last pew and put his head in his hands. "We're never going to find this treasure, fortune, whatever. If it were meant to be, someone would have found it by now," he said. "What if someone already found it, and we're just wasting our time?"

"I don't believe that," countered Davy.

"Me neither," said Billy. "I would have heard about it, and I never heard of anyone finding it."

"I say we keep looking," said Amanda.

"She's right," said Charlie. "We need to figure out what we already know. Does anyone have paper?"

Amanda and Davy started to look around, and then Davy pointed to the pedestal and the book that was lying open. "Hey, what about that?" he said. He walked over and started thumbing through looking for paper while everyone else started shuffling around. Suddenly he said very loudly, "WOW."

"What is it?" cried Pete.

Davy took a piece of paper that had been folded and hidden in the pages of the book. He held it up.

"What's on it?" asked Charlie.

"Come look," said Davy, and he flattened it out over the book. It was a drawing, a replica of the analemma that was on the doors. It had four large stars where the notches were larger. It had numbers written in different locations. On the right of the analemma, near the large part of the eight, at the bottom right was the number 25, followed by the letter F. On the left side, at the bottom was the number 25, followed by an O. Where the eight came together was written 15 A and 1 S. Near the very bottom was 25 D.

"What's all this about?" asked Billy.

"Maybe a secret code," replied Davy. They looked at each other and smiled. Obviously it was the intrigue that made these two so happy. Where there once was animosity, there was now a common bond to find treasure.

"Who could figure this out?" said Pete. "It could be anything."

Pete looked closer at the numbers and letters. He had the vague feeling that he already knew what this meant. His mind couldn't make the connection.

"It's not just anything, it's a clue," said Amanda. She took the paper and sat down in the nearest pew. She brought it very close to her face and said, "Who ever left this here was probably close, very close. We need to figure out what this means." She leaned in even closer.

"Maybe we have to burn it or put a candle up to it or something," said Billy. "I saw that on television, these guys were searching for treasure and they had to heat the paper." He took his flashlight and put it up to the paper. The beam didn't illuminate anything.

"This is regular line paper," said Amanda. "If you put a candle to it, it will just burn up."

"Well, it means something, I just don't know what," said Pete.

Charlie, who was standing near the corner of the church watching said, "Okay, so what do we know?"

"We know the stone fits in all four of the notches," said Pete.

"Okay," said Charlie, "What else?"

"We know the shape is called an analemma," said Amanda.

"True, we know the name of the shape," said Charlie.

"Other people have been looking for it," added Davy.

"Yes, that paper is evidence of that," said Charlie. "Is there anything else?"

"I think the numbers and letters mean something," said Davy.

"Yeah, but what?" said Pete becoming anxious.

They stood in the church looking at each other, and then one-by-one, they started walking around. No one spoke for a long time. Each one seemed to be taking in the clues. Finally, Charlie said, "We can't do any more here. Let's go home and think it out. If you come up with something, bring it to lunch tomorrow." Dejected, they headed for the window. One by one, they climbed out of the dark, moldy church. Inside was a peacefulness that Pete liked. Outside was noisy and hot, he preferred to be inside. He still needed to be alone with his thoughts about Pops.

That night Pete racked his brain. He wanted to ask his grandmother, but decided not to. He wanted to show her he could figure this out, or at least try. He would go to her when he really needed help.

Pete slept fitfully. *In his dreams he was flying in the motion of a figure-eight over the town. He would start at his house, make a swoop to the left, come back and fly over the little church, then swoop right and fly up to Charlie's house, turn around suddenly and swoop around back to the church, and then swoop again to fly over his house. It seemed he did this so many times he was getting ill. The same feeling came to him when you spin in a circle and stop. The world was spinning out of control. The only way he could see himself stopping was to grab on to the weather vane at the top of his house. As he swooped by, he put out his arm, and like a giant rubber band, he was boinged back immediately. He held onto the weather vane, and the wind suddenly picked up. The weather vane changed direction, and was now pointing due north. Pete looked up and could see the North Star, it was directly over his head. In the next moment, he was standing still, and the circumpolar constellations were spinning wildly around the North Star.*

He woke up suddenly and sat straight up in bed. He was in a cold sweat, and out of breath. He looked at the clock, 4:45 a.m. He lay back down and started to slow down his brain. His dream was still so vivid in his mind. Right before it was time to wake up, he fell back to sleep.

Pete was very tired in school that day and found himself staring off into space. At one point, during math, he found himself staring at the globe on the other side of the room. It wasn't the globe he found himself looking at, but the eight-shaped configuration in the middle of the ocean. His eye followed it up and down, just like his dream the night before. Was it that his mind was trying to tell him something, or was it just the fact that he focused on what it really meant. Pete was staring at the analemma.

When math was over, and the class was getting ready for lunch, Pete walked over and picked up the globe. He looked closely at the analemma, and saw months and dates inside the thick outline of the eight-shape. "The letters and numbers are dates," he said.

Amanda was watching him, and came over to stand right behind him. "You're right," she said. She pulled out the piece of paper that they found at the church and laid it on the desk.

Charlie saw them looking at the globe and the paper from the church and came over on the other side. "What is it?"

"The letters and numbers, they're dates on the analemma," said Pete.

"What do they mean?" asked Amanda.

"What if there is only supposed to be one stone, and you place it in the holes on those dates," said Pete.

"What's the first date?" said Charlie.

"October 25," said Pete.

Amanda went over to the calendar that was on the wall. She looked carefully at the month of October and said, "That's a Saturday."

"That's good for us," said Pete. "Because on that Saturday, we should all be in that church."

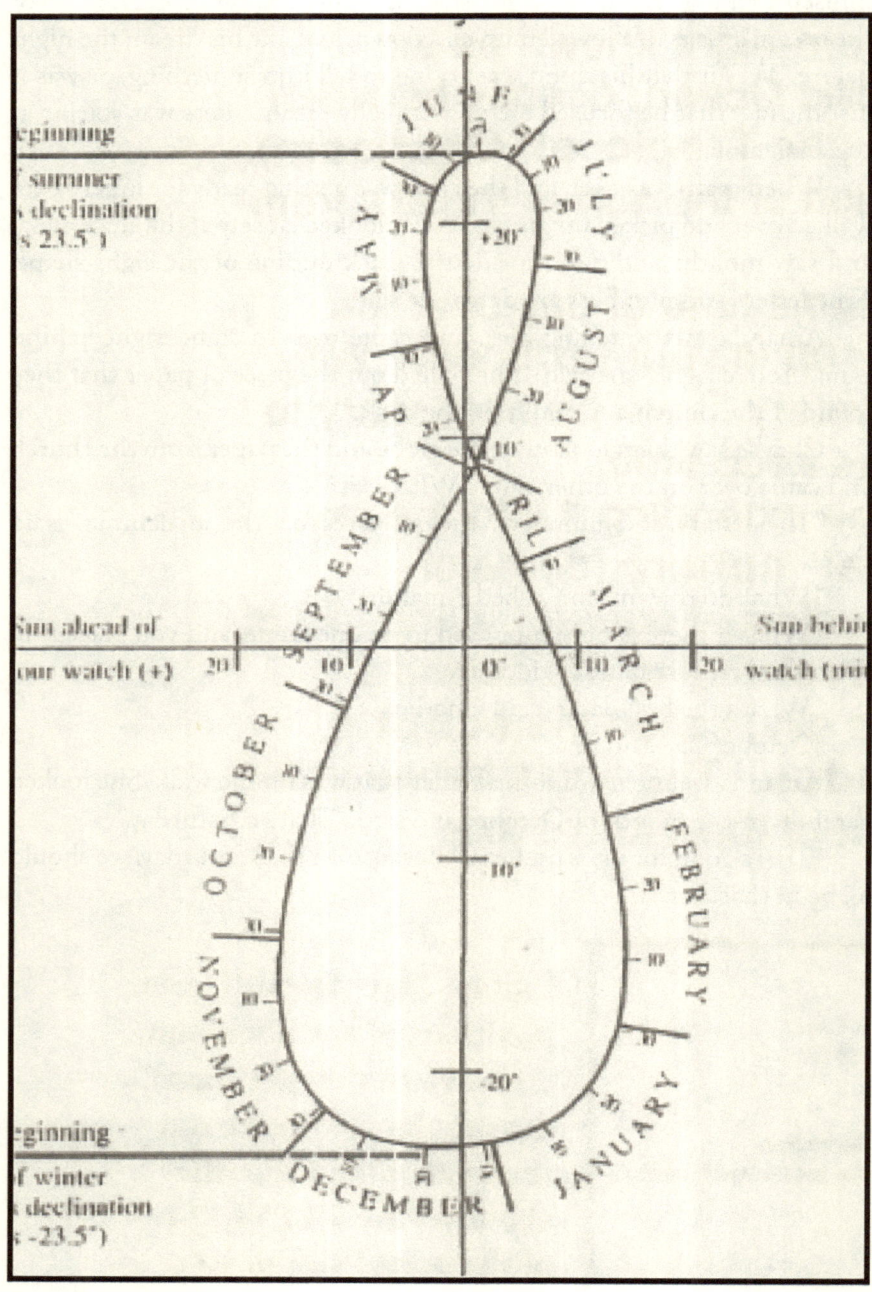

CHAPTER NINETEEN

THE WATCHER IN THE WOODS

At lunch, the sharing of information was top secret. They gathered very closely while Pete, Amanda, and Charlie told of the dates they found.

"October 25, that is really close," said Davy.

"Close, heck, it's this weekend," said Pete.

Billy chewed his food quickly, and in between bites he said, "So what do we do? Do we just put the stone in the notch for that date on the analemma and wait?"

They looked at each other stunned. It was a very good question. Is that all they were going to do?

"What if we have to do something else?" asked Davy. "What if there is a series of little devises that have to be lined up, set up so that the fortune can be revealed?"

A realization that they were probably only halfway there set in. There probably was more that needed to be done, and they were only a week away.

"What do we have for sure?" asked Amanda.

"We have the stone," said Pete. He was thinking of where it was located, and then said, "Maybe the sun clock is important too."

"Why would the sun clock be important?" said Charlie. Then it dawned on him, the stone was is the sun clock, so it must be useful. "Yeah, I think your right," he mused leaning back. He rubbed his hands through his hair. "I bet there are so many things we have to do. How could we be sure we have them all?" He closed his eyes and scratched his head.

Billy scowled at him. "Can't you be positive?"

That was a first, Billy Martin telling someone else to be positive.

"You really want this, don't you?" said Charlie to Billy.

"Yeah I want this," Billy admitted, and pounded his fist on the table. "So let's think about what we need." His sheer determination was enough for the rest of them.

"We really should write down everything we know," said Davy.

He pulled out a piece of paper and started writing.

Sᴜɴ Cʟᴏᴄᴋ
Sᴛᴏɴᴇ, Mᴏʀɢᴀɴɪᴛᴇ
Aɴᴀʟᴇᴍᴍᴀ, Oᴄᴛᴏʙᴇʀ 25

"What else should I write down?" questioned Davy. They looked at each other.

"Do you think the cemetery is part of the key?" asked Amanda.

"It could be," said Billy. "Remember when we found this?" and he pulled the magnifying glass out of his pocket. "We found it near one of the grave stones. Actually, Davy found it." He looked at Davy, "Do you remember where we were when you found it?"

Davy shook his head, "Not really," he said.

The magnifying glass had been cleaned up, and now it looked shiny and clear. "I took it home and cleaned it up. I think it belongs in the cemetery or inside the church somewhere."

Davy quietly suggested, "Maybe we should go there after school and snoop around." This seemed to please everyone, and so the plan was to meet there today after school.

As usual, Pete escorted his little sisters home. Poured them a drink, prepared a them a snack, and then grabbed for the flashlight. Before he left, he asked his grandmother if he was needed for any chores. She was gracious, seeing the flashlight in his hand, and dismissed him to his adventure.

Again, they all met in the back of the church. Everyone had a flashlight and was ready to go. "So what now?" asked Amanda.

"Let me think," said Charlie.

"Maybe we should split up," said Davy. "In the movies, everyone always splits up and looks for clues. It seems that small groups find more clues than everyone together. One time..."

"Can you just be quiet for a moment," retorted Pete.

"No, really, it's a good idea," said Charlie. "We should send two or three outside, and the rest of us should stay inside. We can cover more area that way."

"I'll stay inside," said Billy. He wanted to find out if the magnifying glass in his pocket fit somewhere inside the church.

"I'll stay in with him," said Charlie. He wanted to see if there was anything he missed the night before.

"I'll go outside," said Davy. He was interested in investigating the stone in the eye of the sun emblem on the gate.

"Me too, I'll go outside," said Pete.

"I'll go outside too," said Amanda. She wanted to read the tomb stones, and she was enchanted by the angel statue in the center of the garden. Last time she was out there, she and Davy stopped to look at it. Something was gnawing at her, and she wanted to look again.

When everyone was finished giving their location, the group quickly split up and headed in their separate directions. Billy pulled the magnifying glass out of his pocket. He had it wrapped in Kleenex, which he discarded back into his pocket to use again. He turned on his flashlight and slowly made his way around the room.

Charlie was about to head off into the opposite direction when he saw the magnifying glass. "So what is it you're looking for?" he asked.

Billy turned around and showed him what was in his hand. "There must be a base that this fits into. See if you can find a small square opening somewhere in here," he replied.

They headed off in opposite directions. Slowly a white beam of light illuminated the church in two separate places.

"Look at those stained glass windows," said Billy. "They really sparkle when the flashlight hits them."

"We are not looking at the windows. We don't have too much time. Saturday is only two days away. Concentrate on what you are looking for," said Charlie. His tone was very serious.

Billy began searching for a small square opening. "Charlie, look at those shiny panels on the ceiling, do you think they mean anything?"

Charlie looked up and shined his flashlight on them. The beam bounced back onto the floor. "I dunno, maybe," he replied. "Could you stop talking and keep looking."

The boys slowly crept over every inch of the small church. Meanwhile, outside, the other group was slowly searching the cemetery.

"I think this place is creepy," said Pete.

"You think every place is creepy," replied Davy. "You thought your basement was creepy."

"It is creepy," Pete said. He stopped to see what Davy was looking at. "What is so fascinating about that gate?"

Davy was opening and shutting the gate, and he appeared to have his face so close to the fence, it could possible touch. "There's this small groove with an eight-shape above it. When I close the gate, it disappears. Come look at it."

Amanda and Pete went immediately over to where he was standing. Pete ran his finger over the eight. "You found a clue. I wonder if sunlight comes through here," he said.

Davy stood up and ran his finger over the crystal that was in one of the eyes of the sun emblem on the gate. It didn't look at all like the crystal Pete had. This one looked like cut glass. He looked closely at the other eye which didn't have a crystal. He put his finger on the opening and ran it over the grooves. "Hmmmm," he said quietly.

"Hmmmm, what?" demanded Amanda.

"Look at this eye, feel the opening with your finger," he pointed it out to Amanda. She put her finger on the groove and repeated what she saw Davy do. "I don't get it.""It's not the same as the other. It doesn't feel like there was a brace for a crystal in this eye," he said to her. He moved his head in really close and felt for sharp edges where a bracing could have broken away.

"Look really close," he said to her.

She moved in as he moved out. "I don't see anything," she sighed.

"That's exactly it. I don't think a stone was ever in this eye. I think there is only supposed to be one stone, and the gate has to be closed. Look, the 8 disappears when you close the gate" he said to both of them.

Pete looked very closely at the gate open and closed and how the 8 disappeared, "that could very well be."

Amanda turned on her flashlight. She moved around the yard shining her beam up and down. Davy went back to investigating the gate, and Pete followed behind Amanda.

"What are you looking for?" asked Pete.

Amanda turned around and said, "Last time I was out here, I saw a statue of an angel. Did you see it?"

Pete nodded, "I think so," he moved to avoid the weeds from brushing up against his leg.

Amanda giggled. He was just as squeamish as she was. That was probably why he was out here and not inside the church.

"The angel statue," she continued to talk as she moved into the center of the cemetery, "had its hands up in the air, and seemed once to be holding something." They located the statue, which was almost entirely in tact. It was looking up to heaven, and one hand did look as if it was holding something. Amanda bent down and ran her finger over the hand. She looked closer, and sure enough there was a small 8 inside the palm of the hand, and next to it a small, square opening.

"Oh my," she gasped.

"What is it?" cried Pete who was only inches away.

Davy heard her and made a bee-line to their location. "What did you find?" he asked excitedly.

"It's a small eight inside the hand of the angel," said Amanda.

"No foolin?" exclaimed Davy, and he pushed his way in to see. "I'll be," he said.

"We need to tell the others inside," said Pete. He got up and ran to the open window. Davy and Amanda were still looking at the angel statue when they both heard a noise over in the trees. They looked up immediately to see the branches sway. Davy could swear he saw someone running away, but Amanda was not sure. "Did you see him?" asked Davy.

"I'm not sure. What did you see?"

"Someone was watching us," declared Davy. "He was standing right there. I wonder what he heard."

They stood up and together looked in the direction of the church. "What if there are more people out there, you know, watching us," suggested Amanda.

"I hope not," groaned Davy.

It was not a minute before Billy came running towards them. "So what did you find?" he yelled.

Amanda and Davy looked at each other with a terrified look.

"What's wrong? What is it?" demanded Billy.

"Someone was out there watching us," said Amanda.

Pete and Charlie arrived. They barely heard what Amanda said. "What's going on?" asked Charlie.

Billy couldn't help himself. The mystery of intrigue overcame him and he blurted out, "Someone is watching us."

"How do you know?" asked Pete.

"They were in the woods, right there," replied Amanda, and she pointed to where the branch had moved. "Davy saw him."

Davy insisted, "Yeah, there was someone right there," he pointed to the same spot, "watching us."

"Damnit," said Charlie.

They all looked at him stunned. "What did you say?" asked Pete.

Charlie drove his hands through his hair. He marched around doing circles in the grave yard. He stomped his feet and threw his head down.

"Can you tell us what's going on?" asked Pete.

"I really can't explain it," declared Charlie. "I get this feeling like someone is always watching me. Every time I come up here, or do anything, I feel like I'm being watched."

"By who?" asked Amanda.

"Everyone. My mom, my stepdad, even the teacher," he replied. "What if someone is trying to beat us to the fortune?" He rubbed his face with his hands.

Pete walked over to him and pulled his hands away, "Charlie, it doesn't matter. You see, we will be here on Saturday. If anybody else knows, they will have to be here too. We can catch them at that time."

Charlie stretched his back, and said, "Okay, so what did you find?" He didn't look at Davy, but instead into the woods where the person had been standing.

Davy looked around the group and said, "Amanda found this. It's a square-shaped hole, for something."

Billy, who was holding the magnifying lens tightly in his left hand, pushed it forward, and without even stopping to see if the shapes were the same size, jammed the square end of the handle into the hole. Amazingly, it fit. Everyone's eyes got huge.

"Look here," said Amanda, "there's an eight inside her little palm."

Billy bent his head in and squinted his eyes. "There sure is. I bet that means something," he said.

Davy walked over to the gate. His eyes scanned the woods as he walked. Pete followed him, and so did the rest of the gang.

"What's over here?" asked Billy.

Davy spoke in a hushed voice. "The gate, right here," and he pointed to the gate, "has an eight also. When you close the gate all the way, you can't see it."

Billy got very close and looked at the same eight on the gate. "Wow," he said. Charlie, too, was impressed by the team of kids outside in the grave yard.

"Did you find anything else?" Billy asked.

"Not yet," said Davy, "but we should really keep looking around for more eights." They shook their heads in agreement, and each went a separate direction. They shined their flashlights over every tomb stone in the yard. They looked on the fronts and the backs.

Amanda was reading the stones she had read last time, and she was drawn to the one where the little girl had died within the first year of her life. She felt drawn to this infant, what was her name, there it was, Anita Fairchild. She read it again, and then she saw it. In the A of the first name was a very small eight. You would have missed it if you didn't look closely. "I found it! I found it!" cried Amanda.

Charlie looked directly at her, and put his finger to his lips. "Shhhh" he said. She put her hand over her mouth and whispered, "Its right here."

Davy was the first one there. He knelt down and became personal with the stone. "Yep," he said, "there's an eight right there."

Everyone came over and had a look. "Are there anymore eights?" said Charlie.

"I don't know, but I have to get home for dinner, and then start on all that homework we have tonight," said Pete. "We definitely need to come back tomorrow and see if we can find anything else."

Heads nodded. Billy walked over and pulled the magnifying glass out of the angel's hand. "I'm not leaving this here," he said. They all gave a last look around the yard, and the woods, and then left for the evening.

On the way home, Davy was devising a plan to keep unwanted strangers away from the church on Saturday.

CHAPTER TWENTY

WHAT'S THE BUZZ?

It was Friday, and that day at lunch they all agreed they needed to go back inside the church and look around.

"What if we're missing something inside the church?" said Billy to Charlie.

"We really looked all over inside, and we didn't find anything," he replied.

"We weren't looking for eights. We didn't know anything about the eights when we were looking inside," said Billy. He was right. They had not been back inside since the others found the eights in the grave yard. "We have to go back in there," said Billy.

"I can't go there today," said Davy.

"Why can't you go?" demanded Pete. "We really need you."

"I have to do some work for Mrs. Elders," he said while eating his sandwich. "I'll be there first thing Saturday morning."

Pete shook his head, "Can't you get out of it?" he asked.

"No, not really. By the way, I'm going to make sure no one bothers us on Saturday."

"Oh yeah," said Billy with a mouthful of food, "How you going to do that?"

"You'll see," replied Davy, and he started on his chips.

"That just leaves the four of us, and we need someone to stay outside and keep watch," said Charlie.

"I'll stay outside," announced Pete very quickly. Amanda frowned, she didn't want to go back in the church. But come to think of it, she didn't want to stand watch.

"I don't think I can make it either," said Amanda. She really didn't have an excuse not to go, except she didn't want to.

"Well, that leaves me and Billy to go inside and look around and Pete to stand lookout," said Charlie. "We better comb that place good because the show's on Saturday." Charlie ran a nervous hand through his hair. "Is everyone going to make it on Saturday?"

Heads nodded. No one was going to miss one single minute of Saturday.

After school, Billy and Charlie were waiting for Pete at the church. Billy and Charlie went in, anxious to find clues. Inside the church was cold and damp, much more so than it had been before with the weather changing. Billy and Charlie got right to work looking for the eight-shape.

About an hour later, Pete stuck his head through the open widow and yelled, "Are you guys in there?"

"What took you so long to get here?" asked Charlie coming over to the window.

Pete sighed and said, "My grandma asked me to watch the twins so she and my mom could run an errand. I couldn't say no." Charlie gave Pete a sympathetic look and said, "Okay, you stand guard and let us know if anyone is coming." He moved to the back of the church where Billy was pointing his flashlight at the door.

Pete stood outside near the corner. From here he could see the coming and going from two sides. He liked it here. The woods were on one side, lush and green, and the grave yard was on the other. The graveyard was overgrown, and the church seemed very old. Thick, green moss and ivy were growing on the sides. The roof was made of thick clay shingles, and they appeared to be breaking off in places. The stained glass windows were spectacular with the images of Jesus, Mary and Joseph in vivid bright colors. Pete found himself staring at the windows when the bushes behind him moved. Pete jumped up and looked around. Was it the breeze? He looked at the rest of the trees, and they were all still.

Someone was there, and probably spying on them. Pete drew in a deep breath. What was he going to do? What would a real man do?

The answer was clear. He had to see if anyone was there. Without saying a word, Pete turned on his flashlight and pointed in the direction of the bush. He moved the bush aside, and, nothing. He took a step into the thick of the brush and started to move other bushes around. He took another step, and then something hit him from behind. It knocked him off his feet and he fell forward into the brush. Pete gave a scream and tried to stand up. He struggled in the brush for a few seconds, and then he was on his feet. He looked around to see any movement in the woods. Everything was still. He decided to stand completely still, but this time he crouched down out of site. Who ever knocked him over was probably still there.

He stayed like that for what seemed an eternity, and then he saw 'it'. He couldn't tell if it was a man or woman, but it was definitely a person. It had on a sweatshirt with the hood pulled over its head. He could see the person moving to the window of the church and peering in. Pete watched in amazement. Someone did know what they were up to. How much did this person know? Pete suddenly found a mound of anger welling up inside him. This person, who was spying on them, was probably going to try and take their fortune from them once they found it. He couldn't let that happen. He had to find out who this mystery person was. He had to move, and it had to be quick. Once he gave away his position, the person would be gone down the trail.

He crouched down lower, took a deep breath, and sprang out of the woods in one leap. He was on the person in a split second and knocked him down. He scrambled to his feet and looked straight into the eyes of Mike Sullivan. "What are you doing here?" yelled Pete.

Mike stood up and brushed the grass and leaves off of his sweatshirt. "You little punk. Why I should punch your head in."

Billy and Charlie were at the window. "Who's out there?" yelled Charlie.

"It's Mike Sullivan," yelled Pete. "He's spying on you guys through the window."

"I'm not spying. Just what are you punky kids up to?" said Mike.

Billy backed away from the window, trying to avoid Mike's gaze.

"Billy, is that you in there? What are you doing hanging around with these losers?"

"Uh, we're just looking around in here," said Billy.

"What are you punks looking for?"

"None of your business," snapped Pete.

"Why don't you go home and play Barbie's," Mike taunted Pete.

"We were here first. Why don't you go smoke cigarettes with the other scumbags," retorted Pete.

"I ought to smack you," said Mike pulling his arm back. He turned and looked at the boys standing inside the window. "Yeah, you're looking for the fortune, aren't you? He told me you would be looking for the fortune. What have you found?" demanded Mike to the boys inside the church.

"Nothing," replied Charlie. "Who told you we were looking?" he shot back to Mike.

"None of your business rich boy. I'm watching you," and he pointed to Pete. "I'm watching all of you," and he turned and headed down the trail.

The boys watched until he was out of sight. "At least we know who's spying on us," said Pete with a crooked grin.

"No we don't," Charlie pointed out. "He said someone else told him what we were up to. That kid doesn't have an original thought in his brain." "Are you guys done in there?" said Pete.

Charlie looked at Billy and Billy said, "Yeah, I can't find anything with and eight on it. How about you?"

"Me neither," said Charlie. "Help me get out of here."

They were soon walking down the trail. Charlie was in the rear, and he kept turning around expecting to see someone. Once again he had that strange feeling of someone watching him.

Out from the corner of the church walked a man in a black trench coat. He watched the three boys walk down the trail. He ducked back when the third boy turned around. He was almost seen.

That night, Pete was lying in bed and could not sleep thinking about what might happen tomorrow. When he finally drifted off, *he found himself sitting in the grave yard. The wind was blowing hard over the stones, causing a whistling noise. He saw Davy and Amanda being blown by the wind with their hair moving in slow motion across their faces. Davy was pointing to the gate, and it was flapping hard back-and-forth in the wind. He could tell Davy was trying to walk against the wind to close the gate.*

Off to his left, he saw Amanda standing over the angel in the middle of the garden. She tilted her head, and her hair kept flying in her face. She bent down to touch the hand, and it appeared to move away from her. He could see the magnifying glass in Amanda's hand, and he realized she was trying to put it in the angel's hand, but the angel kept moving her hand away. It was so windy she couldn't catch the hand.

Pete stood up to help her, but couldn't move his body. There was someone holding on to his legs. He looked down and saw Mike Sullivan laughing at him. He had Pete by both legs, and was laughing. His teeth were brown from smoking and hair covered his eyes. Pete tried to kick his feet, but could not move. Mike was much stronger than he, and all he could do was flail his arms wildly.

Pete moaned out loud while he was sleeping. The sound of his own moan woke him up. He sat up in bed and scratched his head. What did all that mean? Was something going to go wrong tomorrow? He rolled over on his stomach and closed his eyes again. It was a long time before he could fall back to sleep.

Pete was very tired at breakfast, and didn't feel like eating. His grandmother was trying to get his attention, but he didn't feel like talking. The dream from the night before still bothered him. Before he left the house, she came up next to him and whispered, "You must consider the 'equation of time.'"

"The equation of what?" asked Pete.

She looked around and said, "Shhhh, the equation of time. If you're going to do what I think your going to do, then you need to either rely on your sun clock, or determine the 'equation of time.'" She walked away, and then turned and winked at him.

Pete left the house confused. He forgot about his dream, but was bothered by this thing called, the 'equation of time'.

Each one showed up around 10:30 a.m. Pete had his sun clock with the stone safely packed inside. He checked it before he left the house. Amanda brought the paper with the analemma written on it. Davy showed up with a large jar with flying insects inside. Pete looked at the jar and said, "Are those hornets?"

"Yep," replied Davy.

"Why do you have a jar of hornets?"

"You'll see," he said, and set the jar down next to a large head stone.

Billy and Charlie came walking down the path, "Are we ready to do this?" said Billy excitedly. Everyone's head nodded.

"I got the stone and sun clock," announced Pete. He decided not to hold it up in case someone was watching.

"I got the paper with the analemma," added Amanda.

"I got the magnifying glass," said Billy, and reached down to his pocket. It was gone. It must have fallen out on his way over.

"Oh no," he said. "It's gone."

"What's gone?" demanded Charlie.

"The magnifying glass, it must have fallen out of my pocket on the way here," cried Billy. "I need to go back and look for it."

"I'll help you," offered Davy, and picked up the jar and followed behind Billy. It was funny how the two of them became friends after all this time.

At that moment, Pete had a horrible thought. Yesterday, Billy acted differently when Mike Sullivan talked to him from outside the church. He seemed reserved and quiet. What if Billy was only hanging around with them to find clues, and he was really working with Mike Sullivan. Pete had to say something.

"Hey, do you remember yesterday when Mike spoke to Billy, how different Billy acted towards him. He acted like they were really friends," Pete said to Charlie.

Amanda was listening intently. Charlie scratched his head, "I don't think that means anything."

"Well, what if he didn't lose the magnifying glass, what if he said that to go get Mike and his gang and force us to show them what we know?" suggested Pete.

"You're just paranoid," said Charlie.

"Am I? I've seen Billy hanging out with Mike and those older guys," said Pete.

"I have too," said Amanda.

Charlie remembered seeing Billy hanging out with Mike that day in the car. It suddenly seemed very real that Billy might be working with another crowd to solve this mystery and just using them. "Maybe we should follow them," suggested Charlie.

The three set out in the same direction that Billy and Davy took a few minutes earlier. Pete, now worried about his little friend, began a fast walk-run combination. Amanda had to run just to keep up.

Billy and Davy were walking slow looking down at their feet. Billy was looking on the left and Davy on the right. Davy didn't hear Pete run up behind him, but they both looked up at the same moment when they heard Amanda and Charlie come running up the trail. "What's going on?" said Davy.

"We just came to help you," said Amanda, out of breath, but relieved to see they were looking for the magnifying glass. All five continued down the path, and then Billy yelled out, "There it is!" They all stopped and looked to where Billy was pointing his finger. The magnifying glass was lying by a big rock in a bunch of tall grasses. The only part visible was the handle. He reached down to pick it up, and a big shoe came out of the bushes and stepped on his hand. It was Mike Sullivan, only this time he wasn't alone. He had four of his friends with him.

Billy had the magnifying glass inside his fist, and wasn't going to let it go. "Get off my hand Mike," he said to the bigger kid.

"What you got there Bill, my boy?" said Mike.

"None of your business," Pete shot back.

"As you can see, puke-breath, I'm not alone," Mike pointed out. "Today it's not four against one. Today the odds are even," and he stared at Pete. Pete glared back, "You and your jerk friends don't scare us," yelled Pete.

Davy turned to Amanda and Charlie and whispered, "Run, right now," and he pointed to the jar. Davy looked at Pete and pointed at the jar. Pete nodded his head in acknowledgment. Davy shook the jar as hard as he could, and that was the cue for the others to run. Davy opened the jar and pointed it toward the older boys. What seemed like a whole nest of angry bees came shooting out of the jar.

"What the ...." was all Mike could say as he was hit in the face with a mob of angry hornets. The boys started to wave their arms wildly around their faces. Davy could see that Billy's hand was now free as Mike defended himself against the hornets, which were clinging to his head and arms. "Quick, run with your head down," Davy yelled to Billy.

The boys ran with their heads tucked down avoiding the swarm of angry hornets. When they were a hundred feet away, they looked back to see the boys had run off in different direction.

"Where did you come up with an idea like that?" asked Billy.

"I just thought we could use some insurance in case somebody wanted to stop us today," explained Davy.

"That was brilliant," admitted Billy, as they walked as fast as they could to catch up with the others.

"Thanks," said Davy, and he ran to keep up.

Before Davy and Billy caught up, Charlie turned around to face Amanda and Pete. "Look, I don't think we should mention what we were talking about earlier," he said.

"Agreed," said Amanda.

"Same," added Pete.

Davy and Billy caught up, and together they made their way back to the cemetery. "We need to hide or something," said Amanda, as she looked around nervously.

"We need to get out of plain sight," said Charlie.

They decided to go inside the church and go over the plan. One by one they slipped into the church, and gathered in the back behind the pews. They were out of sight of anyone passing by the window, and they would be able to hear anyone who came in.

They sat in a small horse-shoe, and Charlie took control. "Davy, you man the gate, make sure it stays closed."

Davy shook his head in agreement.

"Oh, and stay out of sight. Hide behind a head stone, bush, tree, anything. Okay?"

Again, Davy nodded his head and gave Charlie a thumb's up.

"Pete, we need the stone," said Charlie.

Pete carefully took the stone out from behind the sun clock. He looked at the window to make sure no one was watching, and he carefully put the Morganite crystal in the notches on the lower left side of the eight. He pushed it in hard so it would not fall out. Then he sat back down.

"Are you sure you put it in the right spot?" asked Amanda.

"I put it in the spot that was indicated on the analemma for October 25th," replied Pete.

"Billy, you need to put the magnifying glass in the angel's hand," Charlie said. "I'm not putting it in until right before noon. I'm not letting those goons get a hold of it."

"Okay, I agree with that," said Charlie. "What time is it?"

Amanda looked down at her watch, it was 11:30 a.m. "We only have 30 more minutes until noon," she said.

Pete could tell Amanda was nervous, heck they all were. There was something he was forgetting. Something was right in front of him that he was missing. He looked at the analemma on the doors again, and at the placement of the Morganite crystal. What was it that he was missing? Something grandma said, what was it? He looked down at the sun clock he held in his hand. His heart was pounding wildly.

"I'll be outside with the sun clock," said Pete.

Charlie nodded in agreement. "Amanda, can you stay inside with me?"

She shook her head yes. After seeing those big boys, she felt safer inside than out.

"Okay then, its set. We all know what we have to do. If you're outside, you need to hide. We're going to stay hidden in here," Charlie said to the others. "I guess you guys better get going and set up outside. Billy, stay close to the window in case you guys need to tell us something," said Charlie.

Everyone nodded, and Davy gave the motion to zip up their lips to indicate no more talking. The three boys made there way outside. Davy went to the gate and made sure it was closed all the way, and then hid behind a head stone. Billy placed the magnifying glass in the angel's hand, and moved into the bushes by the window. Pete set up the sun clock behind another head stone. Pete looked at Davy and Davy gave Pete his thumb's up. Billy looked over at Pete, and Pete gave Billy a thumb's up. Charlie looked out the window at Billy in the bushes, and Billy gave Charlie a thumb's up. Charlie looked over at Amanda, crouching near the doors, and Charlie gave Amanda a thumb's up. Amanda gave Charlie one back. They were all set and ready to go.

CHAPTER TWENTY - ONE
THE 'EQUATION OF TIME'

The minutes seem to pass like hours. Amanda watched the minute hand slowly move ahead. The anticipation of what might happen filled all five kids. Pete looked up from the sun clock at his surroundings and had an eerie feeling of de-ja-vu. It wasn't just like his dream, but there were strange similarities. Davy was by the fence, and the wind was blowing.

Amanda held her breath as her minute hand moved closer to the twelve.

Pete watched the shadow of the string fall on the twelve, "NOW!" he said.

"Now," yelled Billy to Charlie at the window.

"What?" yelled Amanda, "its not 12:00 noon yet, we still have about 15 more minutes."

Just as she said it, the sun made the crystal in the eye of the sun glow. It produced small prisms in the form of rainbows everywhere. One landed across Davy's face. One hit the magnifying glass and produced a yellow beam aimed directly to the stained glass window with the large eye. From the graveyard, that was all that could be seen. Billy, Pete and Davy all stood up in amazement.

Inside the church, the beam was caught by another magnifying glass in the center of the all-seeing-eye. The beam was directed to the Morganite crystal in the back of the church, and it glowed bright pink. Rings of pink started at the crystal and got steadily larger the farther away. It was spectacular to behold. Charlie stood up to get a better look.

Then in a blinding flash, the rings of pink disappeared and a pink beam aimed towards the ceiling. Charlie and Amanda covered their eyes momentarily, and when they focused again they both saw the beam of light bounce off the reflective panels on the ceiling and come down on one of the pews. They both got up and came slowly to the pew. On the pew where the beam hit was a small, faint 8 carved in the wood. You would have thought it was just a scratch until now.

They looked bug-eyed at each other, and then it was gone. It only lasted 5 or 10 seconds, but it gave the exact location of the 8. "Come in here, come in here now," yelled Charlie to the window. He was afraid to move and take his eyes off the location of the eight. He could hear Billy tell Davy to get the magnifying glass and for both of them to get in there.

Davy jumped up, ran to the angel, carefully grabbed the magnifying glass, glanced at Pete to see if he was coming, and then rushed straight towards the window.

Pete picked up the sun clock and ran after Davy. When he got there, Billy was already inside. He handed Billy the sun clock and then hoisted himself up after Davy. They found Amanda and Charlie standing in the middle of the church, staring at a pew.

"What is it?" asked Billy.

"Why are you guys standing here?" demanded Pete, out of breath.

"Because, this spot is where the beam was aimed. It shot out of the Morganite crystal to the ceiling, and then right down here," said Amanda, and she pointed to the eight. "Can you see it? Can you see the eight?"

Davy put his face close to the pew. It was dark and the shadows in the church made it hard to see. "Yeah, it's there," he said. "What do you think it means?"

"Let me see," said Billy. He ran his finger over the surface and squinted his eyes. "Yep, it's an eight."

Pete put his hand under the pew and ran his fingers across the surface. Charlie did the same thing from the other side. "I don't feel anything," said Pete.

"Me neither," said Charlie.

"Then why here?" said Amanda with a puzzled look on her face. "The beam was aimed directly right here." She put emphasis on the word directly.

Davy, being very small and wiry, got down on the floor and scooted his body under the pew. "Anybody bring a flashlight?" he said.

Charlie turned on his small pen light and handed it to the boy under the pew. "Thanks," said Davy.

They all stood in silence while Davy searched under the pew. Then the pew shook slightly, and Davy said, "I think I got something." The pew shook more, and then a final shake. He let out a long breath of air.

He scooted out from under the pew, and handed a large, heavy object to Charlie. "What is it?" said Pete. They all gathered as closely as they could. Charlie began unwrapping the object which was tightly wound in white cloth. Everyone stood very still and held their breath.

Suddenly a sound came from the front of the church. It seemed to emanate from behind the alter. It wasn't loud, in fact they might not have heard it if they were moving around. Charlie stopped and they all looked at each other. "Let's get out of here," he said to the others. Charlie rewrapped the object and put it under his arm.

"Pete, the crystal," Davy said.

Pete went over to the door and carefully unwedged the crystal from its spot. Very quickly, they were through the window and on the path. Once they were away from the church, they stopped and briefly discussed where they should go. It was decided that Pete's house was the safest place to take the object, and off they ran to see the fortune in the white cloth.

CHAPTER TWENTY - TWO
SOMEONE IN THE SHADOWS

When the last person was through the window, the man in the black trench coat emerged from behind the alter where he had been hiding the entire time. He walked over to the pew and ran his finger over the eight. He bent down and looked under the pew. The mysterious object that the kids removed had been hidden in a small compartment under the pew.

He went back and looked at the door. He examined where the stone had been, and looked at the other locations etched in the eight. He went over to where the beam of light came in from the window. It came in through the all-seeing-eye, in the stained-glass window. He looked up at the reflective panels on the ceiling. "Brilliant," he uttered under his breath. He looked around once more before opening a secret door in the wood panels near the front of the church. The door led directly outside into the grave yard. It was undetectable from inside or out.

He went over to the angel and examined the uplifted arm. He looked closely at the hole where the magnifying glass was not ten minutes ago. He went over to the gate and looked at the crystal in the eye. "Magnificent," he said. He turned and started to examine some of the grave stones. He was looking for the one that had the eight. He couldn't tell from where he was hiding exactly which one it was, but he knew he was in the right place. He studied the stones, but was unable to find the small eight on any of them. He was crouched down in the front of the graveyard when he heard leaves rustling. He turned immediately and hid behind the nearest headstone.

Four large boys emerged from the side of the church. They were fanning out and looking for something. All of them appeared to have red welts on their faces and arms. "I'm gonna kill that little monster when I catch him," said one of the boys.

Just then, the man in the black trench coat stood up. He looked at the boys and said, "You missed them. You missed everything."

"We were on our way to the church when we caught up with that fat kid Billy, and that little punk Davy. We were going to be here, but that little monster released a jar of hornets on us," said Mike Sullivan. "We had planned on being here. It isn't even noon yet. What happened?"

"It was quite amazing, and you missed it," said the man. "You were supposed to be hidden, way before they ever got here. You could have messed up everything. It probably is a good thing you weren't here. At least now I know how it all works," said the man.

"So, what do you want us to do now?" said Mike with his hands dug deep in his pockets.

"We need to get our hands on the stone, the magnifying glass, the piece of paper, and whatever is wrapped in the white cloth," said the man.

"How? I mean who has them?" said one of the other boys.

"It appears as if they all have a different part of the puzzle. The girl has the paper, the boy Pete has the stone, and that little curly headed kid has the magnifying glass. Charlie has the items wrapped in white cloth. I want to know what that is, and I want to know today," said the man in a very angry tone. "I hired you clowns to do a job, and all you do is goof up. So are you going to come through, or do I fire you now?" he demanded.

Mike Sullivan pulled his hands out of his pockets and raised them into the air. "Now settle down old man..." was all he got out before the man in the trench coat was in his face.

"Do not tell me to settle down, is that clear," he spit the words out into Mike's face. Mike backed up three steps and looked down at the ground. "Look at me you fool," said the angry man. "You are one of the poorest excuses for a human I have ever seen. Those five younger kids seem to outwit you constantly. They figured this whole thing out in a matter of weeks. You on the other hand have been looking for this your whole life and where have you gotten?"

Mike shook his head. "I'll tell you where, NO WHERE! You are, with out a doubt, the worst teenager ever. You screw everything up. Everything you touch, everything you try, EVERYTHING, ABSOLUTELY EVERYTHING!" shouted the man in the trench coat. "So the question is, are you going to screw this up too?"

Mike looked at his shoes. He kicked a rock into the woods. He lifted his head and sheepishly said, "No, I can do this."

"I doubt it," said the angry man. "But I really don't have any other options." He walked over to a head stone and was looking at it, when suddenly he saw the small eight hidden in the letter A. He gave a brief smile and took out a small pad of paper and wrote the name of the grave stone down. He looked up to see the other boys watching him. He quickly put the pad of paper back in his pocket and walked to the edge of the grave yard. He turned and looked at them one last time and said, "I expect results immediately." He turned and disappeared into the woods.

"What an asshole," said Mike when he was sure he was gone.

"How are we going to get that stuff?" asked one of the other boys.

"We simply ask, and they will give it to us," said Mike as he cracked his knuckles.

CHAPTER TWENTY - THREE
THE SACRED BOOK

They reached the back porch, and before Pete opened the door, he looked around to see if they were being watched. Everyone then stopped and did the same thing. Pete opened the door and told Davy to take the gang up the back stairs to his bedroom, and be quiet. Pete held the door so it wouldn't slam, and followed.

When they were all in Pete's bedroom, Charlie took out the package, which he had tucked under his shirt. He sat on Pete's bed and unwrapped the white cloth. The object was encased in a black bag with a draw-string. He carefully loosened the draw string and opened the bag. He slid a large book out of the bag. They all looked closer. The book was leather bound with gold leaf pages and words on the front that said 'Holy Bible'.

"What the heck..." started Billy.

Pete cut him off by saying, "maybe the fortune is inside."

Charlie carefully opened the book. The smell was faintly familiar. It was musty and stinky at the same time. The pages were very thin, but unwrinkled. There was fancy handwriting on the front page. It looked like a family tree of sorts. The name Finley Albert Fairchild was at the top. "I wonder if this is his bible," whispered Amanda, pointing to the name.

From the name at the top of the page there were two separate columns leading down, one was Henry Albert Fairchild, and the other was Chester William Fairchild. On the left side under Chester Fairchild, were the names of his children, Marcus, John and Evelyn. Under the center name, John, was Elise Fairchild.

"Look Charlie, your grandmother is in here," exclaimed Amanda. She was clearly taken with the bible and was intrigued by the family tree.

"Where's the fortune?" asked Billy in an irritated tone. He didn't care about some book, much less a bible; he only wanted to know about the fortune, a treasure of sorts.

"Maybe the bible tells us where the treasure is located?" pondered Pete. He looked back down at the book which was still open to the family tree. The branches below the first two names seemed to spread out more, and in the second branch the left side was the name Elders. Davy quickly pointed to the name and said, "I wonder if Mrs. Elders is related to this person."

"Who cares about that old coot," replied Billy.

Davy gave him a sharp look, but said nothing.

On the right side, on the third tier came the name 'James'. Apparently a Michael William James had married a Lillian Mary Fairchild. Pete's eyes grew wide. He pointed to the book, "That's my grandma and grandpa," he said.

Amanda looked at Pete and Charlie, "Well, it appears the two of you are related. "That would make sense, the two largest estates in the town belonged to the founding families."

They looked at the family tree and in one location there was a name with an 'X' through it. Amanda saw it first, "I wonder what that means?" and pointed to the 'X'. "What does it say?" she asked.

Charlie put his nose all the way up to the book and said, "It's really hard to read." He stared for a moment and then said, "It looks like Anita Fairchild."

Amanda blinked and stared off into the corner of the room. "I know that name," she said.

"What do you mean, you know that name?" said Pete.

"I just know that name, I don't know from where, but I remember it from somewhere."

They all looked at her while she scratched her head.

"Amanda, do you know anyone named Fairchild?" asked Charlie.

"No, but I remember that name from somewhere."

Charlie looked back down at the book and said, "lets see what else is in this book." He turned the very thin pages and began to read. "In the beginning..." he read aloud. "This is just the King James version of the Bible. I know, we have one just like it at home. In fact, they look just alike." He thumbed through the rest of the pages carefully, and nothing was inside. No extra papers, no book mark, not even a four leaf clover. "Nothing here," he said, and looked around at the group of kids who were watching him with great interest.

"Let me see that thing," said Billy, and practically snatched it out of Charlie's hands. He put it in his lap and thumbed through it. "What the heck is this?" he said in a negative tone. "All that work and nothing," he said.

"Maybe not," said Amanda, and she gently reached for the Bible. "Maybe it's right in front of us and we haven't figured it out." She took the Bible and put it in her lap and turned to the family tree. "Pete is your grandmother home?" she asked, and turned to look at Pete.

"Yeah, she's probably downstairs. Why?" Pete said, and hung on his last word.

"I think we should show her this Bible, since her name is in it, and ask her what she knows," said Amanda.

Pete looked at her, and then looked around the group. "You all want to do that?" he asked.

They looked at each other and then Davy nodded his head. Charlie did the same, but Billy did not, "If we show her, then she might want to know where we got the Bible," he cautioned.

"I have a feeling she already knows we have it," Pete told him.

"Why would she know we have it?" asked Billy.

"Because she gave me the stone, the Morganite. She knew it was in the back of the sun clock. I have a feeling she might know more."

"She knows about the stone?" asked Amanda.

"Yeah, didn't I tell you guys? I thought I told you guys," said Pete.

"I think you did," agreed Charlie.

"You want me to go get her?" asked Pete

The small group was silent, but the nod of everyone's head spoke volumes.

"Okay, let me go find her," said Pete. He slowly got up and left the room. What are we getting ourselves in for? He thought. He went down the back stairs and searched for the one person he was sure could help, but his stomach was in knots. Why did he feel so bad when there was so much to gain? Was he afraid that what he would learn could hurt her in the end? With apprehension, he searched out his grandmother.

CHAPTER TWENTY - FOUR

A SWEET REUNION

His grandmother was in the living room reading a book. She looked up immediately when he came in the room. It was as if she was sitting there waiting for him.

"Grandma, can I ask you to help me with something?" he said in a soft voice.

"What is it darling?"

He sat down gently next to her and said, "We found something today."

"Is it here right now?"

He didn't tell her what it was, she seemed to already know. "Can I see it?"

"Yeah, we would all like you to take a look at it," he replied.

She got up with amazing speed and grace and said, "Lead the way."

Pete walked briskly up the steps and down the hall, and she was right behind him matching every step. He cautiously opened the door to his room, and four pairs of eyes were watching him walk in. She walked in right behind him. She saw the book lying open on the bed. It was opened to the page of the family tree. She walked over to the bed and sat down. Ever so carefully, she picked up the book. She ran her fingers over the page and closed the book. She held it lovingly in her arms. She took a deep breath and seemed to take in the scent of the book. She put it on her lap and ran her hands over the cover. She ran her right index finger over the gold leaf pages. She sat for a long moment just touching the book. She closed her eyes and pulled the Bible up to her chest and hugged it. The kids watched with amazement and curiosity. What is she doing thought Billy.

She took another deep breath and opened her eyes. She found all of them watching her. Quickly she put the book down in her lap, and opened to the family tree. She put on her glasses that were hanging around her neck and stared down at the page. She ran her finger over the names on the left, and then the right. She did this again and again. Her lips moved as she read silently. After a few minutes, she wiped her eyes and looked up.

"I know these people, these are my people," she said in a reminiscing voice.

"Have you seen this book before?" said Amanda.

"Oh yes," she said, "this was my grandfather's Bible. He used to read scripture to us at night out of it. I would bring it over to him in the living room in this very house. We used to sit and listen to him read. I can remember the gospels and the letters as if it were yesterday." She looked around at their curious faces. "We didn't have television or modern music, all we had back then was radio and books. This was the book of all books," she sighed deeply.

"Then what was it doing in the church?" demanded Billy. He was getting very agitated, and didn't want to hear about old times and reminiscing.

"The old church you are speaking of is our family church, and the secret of our fortune is hidden there," she said.

Pete was amazed to hear these words come out of her mouth. Why hadn't he heard of this before? Why hadn't his father told him of this? "Grandma, why didn't you tell me about the fortune? Does my dad know about this?"

She spoke softly. "Our fortune was in great danger of being squandered by our very own family members, and so it was taken away and carefully hidden. I don't even know where it is located. Your great grandfather, my father, was the only living person who carried the secret, and now he is gone too."

"Why didn't he recover it and give it to you?" insisted Pete.

She shook her small, gray head, "it is too great of a fortune for one person to have. It really belongs to the whole family and should be shared by everyone."

"Do you know where it is located?" asked Davy.

She smiled at Davy, he was such a sweet kid, "No son, I never asked, and therefore I was never told. I always felt rich just knowing there was a fortune to be had."

"So the fortune is real?" asked Billy with wide eyes.

"Oh yes, it's very real, and it's very big," she said.

"Will you help us find it?" asked Charlie.

She looked at him for a long moment and said, "Are you the grandchild of Elise Fairchild?"

Charlie looked around the group and then back at her with a nervous glance, "yes," he gulped.

"You look like your mother," she said. She stood up abruptly holding on to the Bible. "Can I keep this and look at it? The only way I am going to be able to help you is if I can get a closer look at this book."

She looked so authoritative and in control that it was hard to say no. They all nodded their heads yes, so she walked over to the door, and left the room. Pete was aware that she looked very distracted, and obviously had something on her mind.

"She just left with our only clue," said Charlie dismayed.

Suddenly they were all aware that the single and best clue had left the room.

"What are we going to do?" asked Davy.

"We need to think," replied Charlie. "The only thing in that book that could help us is the family tree, and I bet that I have a similar one at my house. I'm going to look for one when I get home."

Amanda was still staring off into space, "If only I could remember where I saw that name," she said. "I want to know why it was crossed out. I got it! That is the name on the head stone, you know the one where the little girl died in her first year."

"Is that the one by the angel?" asked Pete.

"Yeah, why?" said Amanda.

"Didn't you find an 8 on that stone?" asked Pete.

"Yeah! You're right, I did," and she was practically screaming when she said it. "That's the next clue, the head stone."

"There's nothing more we can do here," said Billy. He was deflated because he didn't find the treasure that morning. "What do we do now? We know there's a fortune, but we don't know how to get to it, and our only clue left the room. I can't believe you let her leave with that book," he said in an angry voice.

"That's not true," said Davy. "We need to go back to the cemetery and the church and look for 8's. The treasure is still there; didn't you hear Pete's grandma?"

"There's nothing there, believe me. I went through that whole church and so did Billy. The only real clue is the family tree, and I'm going to go home and find one in my house, and then we need to compare the two. Billy, why don't you come with me," said Charlie.

"Why not?" he said, and lifted himself off the bed.

"Call us when you find it, and bring it over to make a comparison," said Pete.

"Okay," said Charlie, and he and Billy left the room.

Amanda was still racking her brain. "I'm going too," she said.

Davy was the only one left in the room. He looked at Pete and said, "Today was the single coolest day I ever had. Did you see that beam of light go into the church. It was so cool, it like a dream." He sat back on the bed and put his head against the wall. "I don't think your grandma wants to help us find the fortune."

Pete suddenly looked at him and said, "What makes you say that?"

"Did you see the look she gave Charlie? She doesn't like him for some reason. She looked so happy to get the Bible, and if she thinks that the Bible is a treasure, I guarantee she thinks the other treasure, whatever it is, is huge. I don't think she wants us to have it." He didn't look upset in any way. Of all the kids, Davy was really just interested in the intrigue, the mystery of it all. He was so taken by the events that occurred in the last few hours, that he could go his whole life with just that. Pete was amazed at what a great friend Davy was. He was the only one he wanted with him right now. Pete could read people, and Davy seemed truly concerned for Pete's grandmother. How was all of this going to affect her? The knot in his stomach was growing larger.

CHAPTER TWENTY - FIVE

COMING HOME

Charlie had not been too successful with digging up information on his family tree, in fact, all history of the family appeared to be erased. How could it be that with such a fine house, there was no trace of his ancestors? Charlie questioned his mother many times, but with no luck. He went through the junk in the basement, and even wondered into the attic, but he found nothing. How could a family that was so wealthy, and so well known not have any documented history? Charlie was beginning to believe that the lack of information was deliberate.

Billy was beside himself. How could something so cool that happened in the church and cemetery, suddenly vanish in thin air. He went from the verge of insane wealth, back to reality and seventh grade. He was so upset that he stopped engaging in casual conversation, and would only reply in grunts and groans. Charlie found it hard to be around him, and avoided him unless the whole gang was around.

Amanda was completely distracted with the name Anita Fairchild, that she too fell into a single word response stage. She went to the library and looked up the name in the birth records, but found nothing. She couldn't shake the feeling that the name had some real important significance. It bothered her both day and night.

Pete was worried about how this Bible could affect his grandmother. He felt bad about getting her involved. He kept a watchful eye on her during meals, before and after school, but she didn't waver in her ability to take care of him and his sisters. He wished his father were here. He had almost completely forgotten what it was like to have him around. A stray letter or phone call from his father was all he had.

Davy was the only one out of the group who stayed focused. He had made the connection about the 'equation of time'. He figured out that solar noon occurred 15 minutes before clock noon. It was probably a very good thing that solar noon happened so early because Mike Sullivan and his friends were heading back to the church. As it was, they were already gone before clock noon occurred. Any later, and it may have turned out differently. He was already planning the next time they would put the crystal in the analemma on the door to see what would happen. He studied the map of the analemma, and they would be ready to go on or about December 25th. According to the analemma, clock noon and solar noon would be exactly the same on that day. The 'equation of time' would zero. He was anxious to see if the events would happen the same way it did the first time. The most important factor now was to make sure that no one, absolutely no one, except the five knew this information. In fact, it really bothered him that Mike Sullivan had found out they were at the church that day. They must have been over heard talking about it in the cemetery. This could not happen again.

That Monday at school, they all put their heads together during lunch. Davy had a piece of paper and was taking notes. Pete started the conversations, "My grandma hasn't said anything to me about the family tree, Charlie did you find anything out?"

Charlie groaned, "No. It seemed we just appeared out of nowhere. No one knows anything."

"How can your mom and your grandma not know anything? My grandma knows all of the history of this town." said Pete.

"My mom doesn't care about her family history, and my grandma is in a home for old people. My mom visits her, but doesn't bring us anymore."

"What I think we should do is go back to the cemetery and look at the grave stone that has the 8. What was the name?" asked Pete.

"Anita Fairchild, and I think that is a great idea," said Amanda.

"Can everyone go after school?" asked Billy.

"I can," said Davy.

"Me too," said Pete.

"Me three," chuckled Amanda.

"I'm in," said Charlie.

Davy leaned in and whispered, "We need to make sure no one knows what we are doing. Somehow, those other boys found out what we were doing, and almost beat us there."

The gang nodded in agreement. "Okay, tell no one and bring a flashlight."

The plan was set, everyone was to go home, keep their destination a secret, and get to the cemetery as quickly as possible. Pete had it in his mind that he would be there first. He got his sisters off the bus, and as fast as they could possible move their little legs, they were at home. Pete herded them in the back door, careful not to let it slam shut, and had them in the kitchen, a snack in their hands and water in their cups. He put his backpack in the corner by the steps, and grabbed the flashlight from the cabinet. He was almost out the door, when he heard his name called from the living room. He made a heavy sigh, dropping the flashlight to his side. "What?" he groaned.

Grandma appeared in the hallway with her hands held out in front of her. "Pete, girls, I have a wonderful surprise for you. Come in here, quickly." She turned and was swiftly down the hall. The girls followed in great anticipation, but Pete sauntered swinging the flashlight next to him. What is this he wondered?

When he entered the room, there was a strange man sitting in the chair by the picture window. He immediately stood up and held out his arms. Pete squinted his eyes and said, "Dad, is that you?" The man, who looked old and frail, nodded his head 'yes'.

Pete took two very large steps and was in the arms of the man. The little girls too, ran to him. In one swoop, the man gathered all three children, and they hugged. The hug was enormous. Pete's mom and grandma joined the hug, and then there were kisses, more hugs, and tears. Pete was overcome with joy, that he had tears streaming down his face. He forgot all about the kids at the cemetery until he felt the weight of the flashlight in his hand. He swallowed hard, and made the decision to stay home and be with his family. He put the flashlight down, and hugged his father again.

Meanwhile, at the cemetery, the others were waiting. "So where's Pete?" said Billy to Davy. Davy shrugged his shoulders and said, "I thought he would be here by now."

"I guess he's not coming," said Billy. He still had hard feelings towards Pete.

"Let's start looking around," suggested Amanda.

The four took their flashlights and went over to the head stone that said Anita Fairchild. They found the small 8 hidden in the letter A. They took turns rubbing their fingers over the 8. Charlie and Davy combed every inch of the head stone, and found nothing else.

"Let's look at the angel again," said Amanda, and so everyone trotted over to the angel. They examined the angel, and saw the 8 in the palm along with the square hole for the magnifying glass. Davy sat down in the middle of the cemetery and put his head in his hands. "You know what I think?" he said.

"No, we do not know what you think, tell us," said Billy in an impatient voice.

Davy shook his head and swatted a bug out of his face, "I think we need to find out what is in that grave."

"We can't do that," said Charlie.

"Why not, I was thinking we should see what is in that grave too," exclaimed Billy.

"It's called grave robbery, and it's illegal," said Charlie.

"But that's our next clue," said Davy. "How are we going to get to it if we don't dig it up."

"We can't dig it up, and that's that," said Amanda. "Charlie's right, we could get into really big trouble if we just start digging in a cemetery."

"Well, then what are we going to do?" said Billy.

"We'll think of something," said Amanda. She put her hands on her hips and turned slowly around looking at the cemetery. As she was turning, she saw a bush move. She stopped suddenly and said in a low voice, "Someone is watching us."

"Where?" said Billy and Charlie at the same time. They moved next to her and looked where she was staring into the woods. Again, the bushes moved. "You see them?" she said.

"I see the bushes move," whispered Charlie.

"Lets get out of here," said Amanda. She didn't like Mike Sullivan, and was afraid that he was going to hurt one of them.

Charlie, seeing the fear in Amanda's face said, "Yeah, we should go."

Davy stood up and brushed off the back of his pants and followed the others already making their way out of the cemetery. When they got to the path that led away from the church, they were surprised by a person jumping in front of them in the path. It was Mike Sullivan, and he was immediately surrounded by four of his friends. They moved in closer to surround the small group of kids.

"So what have you been looking for?" said Mike.

Suddenly very uncomfortable, Billy took a step backwards.

"We were just looking around," said Charlie in his most casual voice.

"What was in the package?" he demanded, and got in Charlie's face.

Charlie remained very calm and said, "What are you talking about?"

Mike pushed Charlie backward, and he hit Billy square in the chest as he lost his balance. "Hey, knock it off," said Billy.

"What are you going to do about it tough guy?" he said to Billy.

Billy was getting very upset, it was clear by the redness that was developing in his face. "Get out of our way," he huffed back.

Mike moved into the middle of the group and got close to Amanda. He took a piece of her hair and flipped it up. She didn't move a muscle. He turned and put his face up to Davy's. He put his hand on Davy's head and ruffled his curls. He stepped in very close to Charlie and traced the fancy emblem on Charlie's shirt, but he stopped and stared at Billy. He put his sweaty face right into Billy's and said, "So where have you been, Billy? I haven't seen you around lately. You go and get new friends? Aren't we good enough for you?" He poked him in the chest and then pushed him hard. The other boys laughed along.

"Stop it!" yelled Amanda. "Leave us alone. We haven't done anything to you."

"Just being here bothers me." He turned to face her, "You have something I want, and I want it now." He stepped back into the middle of the group and turned around slowly so that all of them could see him clearly. "I want what you took from the church Saturday, and I want it now!"

It was Amanda who spoke, "We don't have it anymore."

Billy piped in, "It was just a dumb Bible. It didn't have anything in it."

"Someone must have left it in the church, it is a Bible and that's a church. We gave it away, and that is why we are here looking for more clues. It didn't tell us anything," replied Charlie.

Davy just kept quiet. Being the smallest member of the group, he was used to being picked on.

Mike stood there getting mad and turning slowly in circles. He squinted his eyes as he looked at each one of them. He tried to look as intimidating as he could. He wasn't going to be successful here, but he wanted them to be afraid of him the next time they saw him. He stopped when his gaze fell on Davy and punched him hard on the shoulder. "That's for the hornets, punk." Davy fell backwards into the bushes. He had the feeling he was going to get it. It hurt very badly, but he wasn't going to cry.

Amanda came over and helped him get up. She gave Mike Sullivan her meanest look. "Only bullies pick on someone smaller than themselves."

One of the older boys was tired of watching and said, "I'm leaving, this is stupid." He turned and left. Two other boys followed behind him. Only one stayed, but he looked around as if he might follow also. He said, "Come on, let's get out of here."

Mike put his face close to Amanda and said, "You're next."

She back away and scowled at him.

After they left, Amanda let out a huge breath of air. "Who do they think they are?"

"They are looking for the treasure too," said Charlie. "We really have to watch what we do."

"What if they dig up the grave and get the clue?" asked Billy.

This upset Charlie and Davy. Both of them looked at Billy with concern. There was no way they would ever agree to digging up a grave.

"I don't care about the treasure. I just want to get out of here. I'm not coming back if that creep is going to be here," said Amanda.

She started walking on the path that led to the street, she wanted to get home. Davy was right behind her and was thinking of something to say. He didn't want her to leave the group.

When they got to the street, he caught up to walk beside her. Usually very talkative, he was at a loss for words. When she came to his street, he stopped and said, "Hey, are you Okay?"

She was surprised, "He punched you, not me. Are you Okay?"

"Yeah, I'm fine. Don't worry about me. I don't want you to leave us, I mean the group. Are you going to hang out with us again?"

She smiled and sighed, "Yeah, I'll hang out. I just don't like that guy."

"Me neither, but he isn't going to stop me." Davy smiled, happy to hear she wasn't going to leave them. "See ya tomorrow," he waved and walked off.

"He scares me," she said in a low voice, and watched him walk away.

CHAPTER TWENTY - SIX
DIGGING-UP THE PAST

The next morning Pete came into the kitchen and was surprised to see his dad sitting at the table drinking coffee. He seemed so different to Pete. He looked older with graying hair and wrinkles. His eyes seemed sad and always worried. Pete couldn't believe he was really home. "Are you going back?" asked Pete.

"No, I'm not going abroad any more. I'm going to take a job in the states. We may be moving to New York or Chicago soon," said his father.

Pete was immediately alarmed. "We can't leave," he blurted out.

"Why, you like it here?" asked his father.

Pete, stunned by the thought of leaving, took a moment to collect his thoughts. "I, I, don't want to leave grandma. Who's going to be here if she needs help?" Pete left out the part about the treasure and his friends.

"I'm still talking to the networks, and it may take a little while. But I want you to be ready to move. I have missed all of you so much. We are going to stay together." He drank his coffee, and looked sternly at Pete, "So how are your grades? Are you doing well in school?"

Pete, still stunned by the announcement of moving, looked up and responded, "Yeah, I make all A's."

"Do you help out around the house? What kind of chores do you do?"

Again, Pete answered with very little enthusiasm, "I take out the trash. I get the twins to school and back everyday. I watch the twins when mom and grandma are out."

"Is that it? Boy, when I was your age I had a paper route. I got up at five in the morning everyday to deliver papers. I had serious chores, and I was responsible." He took another drink of his coffee.

Pete suddenly felt small, like he was being lectured for not being good. He looked down at the table and let out all the air in his lungs. Happiness had turned to disappointment.

Pete's grandma sat down at the table and put her hand over Pete's and spoke up, "Pete is a fine young man. He does a lot more than he says. You should be very proud of him." She squeezed his hand and spoke again. "He has become very dependable, and I enjoy his company."

Pete realized she had said the last part for him. She was letting him know that she really was glad he was with her. Pete looked at her and smiled. She bent over and kissed his forehead. He didn't want to leave her.

At school that day, Pete let his mood show as he entered the classroom. Charlie and Amanda picked up on it immediately. "What wrong with you?" said Amanda.

Pete had a hard time forming a sentence. "My dad came home last night."

"Wow, I would think that was good news," said Charlie.

"Yeah, it's good to have him back, but he wants to move," replied Pete.

Charlie and Amanda stared at him. "Why? What's wrong with Preston?" asked Amanda.

"Too small, I suppose. He wants a big city," said Pete.

Caught up in Pete's news, they forgot to tell him about running into Mike Sullivan at the cemetery. It wasn't until lunch time that Pete heard the story.

"So do we dig up the grave?" asked Billy excitedly.

"You know we can't do that," replied Amanda.

"Maybe we should," said Charlie.

Amanda turned and gave Charlie a stunned look. "You changed your mind?"

"If we don't, Mike Sullivan will. What ever is there will be gone, and we will lose the chance to find it."

Amanda changed her gaze to Pete, a voice of sound reason, and said, "What do you think Pete?"

Pete was listening, but couldn't shake the conversation he had with his dad out of his head. "Yeah, let's do it."

"So throw caution to the wind?" asked Amanda. "What if we get caught?"

It didn't seem like it was going to be an issue. Amanda shook her head. She was the only one who didn't like the idea. "I'm not going to be there when you do it."

"Okay," said Billy casually.

Amanda was immediately angry. "So your going to leave me out?" she asked the whole group. Charlie looked down, and Billy shook his head, "Yeah, probably."

Her face started to turn red, "You're not going to leave me out. I have to be there to make sure you guys don't get in trouble."

"Really, if you want to stay out of it, it's okay," said Charlie.

"No way, I'm coming," she argued.

"Girls," said Davy.

Amanda huffed and ate the rest of her lunch without speaking.

Before leaving school that day, they all agreed that they would meet that evening. They would gather after the sun set, and dig up the grave. They would need flashlights, shovels, and total secrecy. They couldn't tell anyone where they were going.

On the bus ride home, Amanda avoided Pete. Pete didn't understand why she was mad, and finally asked her. "So why aren't you talking to me?"

"You all seem to think its okay to dig up graves. Are you all mad?"

Pete tried not to smile, "Amanda if you feel uncomfortable, then maybe you should stay home." He was trying to calm her fears.

"So that's your answer, leave me out. No way, I'm coming too. Who's going to keep you all out of trouble?" Her face was turning red again.

Pete realized there wasn't much he could say to her. He turned and faced the front of the bus and waited for his turn to get off. He didn't want to dig up the grave, but he didn't see any other way to find the clue.

Anticipation and anxiety plagued Amanda the rest of the day. She didn't want to dig up a grave, she was afraid of getting caught, but she was more afraid of being left out. She couldn't stand the thought of the rest of group finding a clue, or the treasure, and not being there. She had to be part of the action.

When Pete got home, his grandma was sitting alone in the kitchen. Pete went over to her and gave her a squeeze. He appreciated her kind words earlier.

Pete's grandma was very happy to have her son back home with his family, but disappointed at the thought of them moving away. She enjoyed the house filled with the sound of children. She knew it wasn't going to last forever, and now she had come to realize it.

She had spent many nights looking at the family tree inside the Bible. She wasn't getting anywhere. The name Anita Fairchild wasn't familiar with her. She couldn't remember her cousin Elise giving birth to that little girl. Did her cousin have a baby and it died? She was angry with Elise for stealing her boyfriend, and she didn't talk to her for a long time, but she still would have heard about this. She made up her mind that she would confront her in person. That same morning after the kids were off to school, she grabbed the Bible and left the house. The drive to the nursing home was short. She barely had time to think about what she was going to say. Their families had not been close, and she hadn't spoken to Elise in over 40 years. Would her cousin even recognize her?

When she entered the building, there was a hostess station ten feet from the door. She cautiously walked to the counter and asked for the room of Elise Fairchild, and held her breath. A moment of panic struck her. What if they wouldn't let her in to see her cousin? What would she do then? She had to know now about the baby. It was burning in her mind.

The receptionist came back to her and said, "Mrs. Fairchild is in room 216. You go down this hallway, up the stairs and make a right. Her room is down the hallway on the right." She gave a warm smile.

The elderly woman walked slowly down the hallway. Now that she was on her way, she was not sure what she was going to say.

She peered inside the room. It was quiet and seemed empty. The fake smell of flowery air freshener filled the air. She took a step in, and looked around. Two twin beds were in the room. It was larger than she thought it would be. The large picture window was open and framed with stripped curtains. She saw an old woman sitting in a chair facing the window. Both beds were neatly made, and the curtains were softly blowing. She came up to the person sitting in the chair. She was asleep and her head was tilted off to the side. She took this moment to study the woman in the chair. Her hair was pulled back into a neat bun, and her face was adorned with make up. She looked angelic sleeping. Even in her old age, Elise Fairchild was beautiful. She stood there for a moment, and then the old woman in the chair jumped. She noticed the other woman in the room and gave a little scream. "Who are you?" She looked closer and said, "My God, Mary is that you?" She squinted her eyes and said, "Goodness, you're old."

"So are you. You want me to give you a mirror so you can see how you look?" said Mary. She moved around to the other chair that was in front of the window, and slowly sat down.

"I haven't seen you in years. What brings you here now?" said Elise. Obviously no love was lost between these two women.

Mary sat straight up; she held up the Bible in both hands and said, "I have a question to ask you."

Elise saw the Bible and her curiosity was peeked, "Did you come to save my soul?"

Mary gave a grunt of a laugh and replied, "I'm not sure that I, and this Bible would be enough to complete that task. No, I came to ask you a question." She opened the Bible to the family tree, and turned the book so that Elise could see it. "Tell me about Anita? Did you lose a baby?"

Elise stopped rocking and stared off into the open window. She acted as if she didn't hear the question.

Mary waited for a few minutes and then leaned in closer. "Did you hear me you old coot? I asked you about the baby named Anita Fairchild, did you have this child or not?"

Elise suddenly turned her head and anger shot out of her eyes. "Don't you think I don't know what everyone says behind my back? I loved him and I wanted to keep him. The only way I could think of was to get pregnant, and then he would have to marry me. The thing was, I never got pregnant, and I tried." Her eyes welled up and she began to sob. "So I told him I was pregnant anyway. I stuffed a small pillow under my shirt." She made a gesture with her hands of stuffing under her shirt. "He believed me. He said we would be married after the baby was born. I had to produce a baby, or I was going to lose him." She squeezed the tissue she had in her hand. "I told my papa that I wanted him, I told my papa that if I didn't have him, I was going to go crazy. Papa loved me so much. It was his idea to say the baby died."

She took a deep breath and sobbed loudly. Her body heaved as she cried. For a couple of minutes Mary watched, and then she took her hand and put it gently over Elise's.

Elise pulled it away and glared at her. "You didn't care. You stopped coming around. I wanted to tell you. I wanted you to tell me what to do, how to get him back. I needed you and you left me."

Mary sat stunned at the information she was hearing. "You pushed me away when you took Michael from me. You knew I loved him, and you took him anyway."

"Don't be ridiculous, I never wanted that country bumpkin, he wanted me."

Mary shook her head in disagreement, "That's not what he said."

"Of course you believed him, you wanted to."

"Tell me about the baby in the grave," said Mary.

"There is no baby in the grave, there's just a head stone marking a grave of a baby I wanted to have. I lost him anyway." She stared vacantly out the window.

Mary was suddenly overcome with a burning question, "So who was this man that wouldn't have you?"

Elise, again, glared at Mary with anger, "Why do you care? You couldn't have ever had him. You were plain and simple, and he was magnificent. I know he really wanted me. He told me how beautiful I was. He told me I was the only woman he would ever love."

Intrigued, and feeling slightly angry, Mary leaned in and said, "But you didn't get him, did you?"

Tears streamed down Elise's face. "Don't you think I don't know that. I think about him every day. He was the only man I really loved."

Again, Mary touched Elise's hand. This time she didn't pull it away. "I'm sorry for you. You really did get everything you wanted. It must have been hard for you not getting him."

"Are you mocking me?" Elise raised her voice.

Mary's eyes were kind, and her hand was holding on to Elise's, even though she tried to pull away. "No dear, even after all these years, I still have a soft spot for you."

They sat there, holding hands for awhile, and then Elise pulled her hand away and said, "What do you really want? Are you here to get money?"

Mary stood up, "I have what I came for. I'm going home to my family," and even as she said it, she knew very soon her family may be leaving.

"You'll be in here soon, your family will do the same thing as mine. They'll drop you off and never come back."

Mary started walking out of the room. She could hear Elise yelling at her as she continued down the hallway. "Come back here, I'm not done with you."

"I'm done with you," whispered Mary under her breath as she left the building.

CHAPTER TWENTY - SEVEN
THE SECRET OF THE HEADSTONE

Pete sat at the dinner table. He ate what he took, which wasn't much, and when he was done, he stared off into space. He heard the others talking, he saw his little sister giggle at something his dad had said, he saw his mother looking lovingly at his father and then he saw his grandmother. She was looking at him, trying to make eye contact. He snapped out of his trance and looked directly at her.

"I have to tell you something," she said softly for only him to hear.

Pete nodded his head, and started to pay attention to the people around him.

He took out the garbage without being asked. He swept the floor and straightened the chairs around the table. He kept a watchful eye on the time. It was getting dark and he needed to get ready. He removed the flashlight without being noticed. He was heading for the door to get a shovel, when he saw his grandmother wave to him. She moved into the laundry room. When Pete entered, she quietly shut the door.

"I found out something today," she said in a hushed voice.

"Is it about the Bible?" Pete also spoke very softly.

"Yes, I found out from the source herself that the grave with Anita Fairchild is empty. There is no one buried there. It's empty."

Pete had a puzzled look on his face, "Who's the source?"

Grandma took a deep breath, "My cousin, Elise Fairchild, Charlie's grandmother. She told a whopper of a lie, and it had to end with an empty grave. I guess that is why the name is crossed out on the family tree."

"Do you suppose the treasure may be buried there?"

"Why do you think the treasure is there?" asked Grandma.

Pete looked at her in the eye. "Grandma, there's an 8 in the A on the headstone. Do you suppose the treasure could be there?"

Grandma now looked puzzled, "Pete, is that what you think?"

"Maybe, yeah. I think it could be buried there."

"Are you going out there?" she asked.

"Yeah, me and the others are going out there tonight."

A noise came from the kitchen and the door suddenly opened. "When you wash colors, you use cold water," said Grandma. "Pete was asking if he could help with the laundry."

Pete's father stood in the open door. "It's about time he had more responsibility. Pete when you're done, can I see you?" Pete nodded, and he turned and left the small room.

Pete looked concerned. "I won't tell, just don't get caught," she whispered in his ear.

Pete went to the living room where his father was sitting. "I'm glad to see you doing more around here. I was hoping we could talk later, after you finished your homework. We need to catch up. I want to know more about you."

"Sure, that would be great," said Pete trying to avoid making one word responses. "I need to get started on my homework."

"Of course, I will see you later then."

Pete turned and left the room. He found a small shovel in the shed and put it under his shirt to hide it the best he could, and then ran swiftly to the grave yard.

When he got there, the others were gathered around the grave. It looked as if they had already been digging. "Did you find anything?" he said with excitement.

Charlie moved aside so that Pete could see the grave. "It was like this when we got here."

"What do you mean?" asked Pete.

"Billy and Davy got here first, and Amanda came shortly after. It was already dug up."

Pete looked at the area. There were a sloppy series of holes, with mud slung everywhere. It looked as if a central hole was dug, and then many small holes around the larger one.

"It doesn't look like they found anything," said Charlie. "They dug so many holes, and nothing."

"What if they did find it?" said Billy. "How are we every going to really know?"

Davy just shook his head.

"I think Charlie's right. I think they dug all these holes and couldn't find anything," replied Pete. He suddenly remembered what his grandma told him before he left the house. "There's no body here."

"What?" said Amanda turning to face him.

"My grandma said that she visited your grandma in the nursing home, and she told her that there really is no one buried here."

Charlie was stunned. "She visited my grandma today? She told her that there was no body?"

"Yeah, that's what she said."

"Then this would have been a perfect place to hide the treasure. What about the 8 on the headstone?" questioned Billy.

"I was thinking the same thing, and now there are all the holes," said Pete.

Consumed in the conversation, no one heard the person approach. "What the heck is going on here?"

They all turned immediately to see a man standing behind them. It was Pete's dad. He moved in closer to take a look at the grave. He looked at the shovels the boys were holding and began to yell. "What have you done?"

He looked at the faces in the small group, and then his gaze rested on Pete. "I can not believe my eyes. It is a crime to dip up a grave. Pete, what have you done?"

"We didn't do this?" said Davy. "It was like this when we got here."

"Oh really, then why the shovels?" He stepped into the center of the circle, and faced Pete. Davy stepped back, hit his foot on the shovel and fell backwards just missing the headstone. He landed between two piles of dirt. He pushed on the headstone to get up, and when it did, he felt a sharp metal edge. "Ouch," he said. He pulled his hand away to see drops of blood already forming. Oh great, he thought. He looked at his hand, and then it hit him. What could be sharp on a headstone? He slowly moved his body to the back of the headstone where Pete's dad couldn't see him. He got very close and ran his fingers at the base of the stone, and there it was. Very close to the ground was a small, metal plate almost flush with the stone.

Billy was watching Davy. He could see what he was doing. The only other person who could see Davy was Pete, and he was getting yelled at. Billy doubted if Pete was paying attention to Davy.

"I thought you were doing homework. I looked upstairs and you weren't there. How do you think I felt when I asked to have a talk with you, and you disappear? If it weren't for your grandma, I wouldn't have found you."

"Grandma told you I was out here?" asked Pete.

"No, how would she know that? She told me you were over at Charlie's house. I called there and Charlie's stepdad told me that you sometimes hang out here."

Charlie was immediately surprised. He didn't tell his stepdad where he was going. He never mentioned going to the cemetery, the old church, or anywhere to him. They rarely spoke.

Meanwhile, behind the headstone, Davy put his fingernails under the side of the plate and tried to pry it loose from the stone. He ran his nails up and down the plate, and then gave another yank. To his surprise, the plate came loose, and slid outward. Davy almost fell backward losing his balance. The plate was small, only about the size of a credit card. Inside the flat plate was a small, yellow piece of paper. Davy took it out and put it in his pocket. He ran his fingers on the bottom of the plate to see if there was anything else in the secret compartment. It was empty. He pushed it back in and poked his head around the headstone. He saw that Billy was watching him. He nodded to Billy, and Billy nodded back.

"What did you find here?" asked Pete's dad.

"We didn't find anything here. It was like this when we arrived," said Amanda.

"How do I know you are telling the truth?" demanded Pete's dad.

"We don't have anything, look at us. No one is holding anything," said Pete.

Pete's dad took a deep breath and let it out with disgust. "Where is the body that should be here?"

"There was never a body here. It's an empty grave," said Pete.

"An empty grave? How do you know that?"

Pete realized that by telling his dad the truth, he would implicate his grandma. "It just is," said Pete.

"I want answers and I want them now!" yelled the man.

"It was like this when we got here," said Charlie.

"I've had it, Pete let's go," and he reached for his son's arm. He pulled him abruptly, and Pete moved with him. They left rather quickly, and they could hear Pete's dad yelling at him as they went.

"Let's get out of here too," said Davy.

They picked up their shovels, and left the cemetery. Davy and Billy walked together on the way home. "I found the clue. It was in a secret compartment on the side of the headstone."

"I saw you take it out, lets look at it when we get to your house," said Billy.

They went straight to Davy's room. It was the first time Billy had been in Davy's house. It was a lot like his, same floor plan, same hard wood floors as his house. Inside Davy's room, they looked at the small, yellow piece of paper. On one side were the letters S H E, and on the other were letters written in groups:

TT

LA   RM

BG

"What does this mean?" said Billy.

"Just a minute. Clues take time to figure out," replied Davy and he stared at paper. He turned it over and stared at it again. "What ever we do, we keep this a secret."

"Are you going to tell Pete?" asked Billy.

Davy looked up at Billy, "Yeah and you're going to tell Charlie. And one of us will tell Amanda. We just need to keep it a secret among us."

"Okay," said Billy. He pointed his finger at the small piece of paper, "What is this supposed to mean?"

Davy was staring at the small paper, "I don't know, but we'll figure it out."

"I hope so," sighed Billy.

Chapter Twenty - Eight
SHE

To what end will this lead, thought Pete. His father was raving mad. He was grounded for the next month, and he was to write a formal essay explaining what he was doing in the cemetery that night. The whole family was alarmed by the fuss his father made when they came into the house. He was ordered to go to his room where he would have to stay before and after school, unless he is doing chores. His father continued to yell even after he was in his room and the door was shut. By the time he would be ungrounded, they would probably move. He didn't think he was ever going to hang out with his friends ever again.

Davy told Pete what he found. Pete told Amanda, and Billy told Charlie. The group was sworn to secrecy, not even Pete's grandma was to know the clue. No one ventured back to the cemetery, afraid they would be questioned about the disrupted grave. No one could figure out the mystery to the new clue. What did SHE mean? What were the initials on the other side of the paper?

Amanda guessed that the T stood for top, the B for Bottom, the L for left and the R for right. Everyone liked this idea. Davy suggested the second M was for Morganite, since that was the stone they already had. Everyone agreed on that too, but the rest was a mystery.

Days were growing shorter, and Pete noticed this as he followed the dates on the analemma. Soon winter would begin and they would miss their next opportunity to put the stone in the hole at the church. In fact, all of them were very aware that time was running out.

Pete couldn't leave his room after school, and so Davy hung around with Billy. Odd as it seemed, the two got along quite well. Davy was helping Mrs. Elders with the chores around the house. Almost everyday after school he would rake, mow or help clean the garage. Billy had gone over once and helped, at the shock of Mrs. Elders. Davy took the opportunity to explain to Mrs. Elders the new friendship that had formed between him and Billy. She could hardly believe it, and didn't want to give Billy too much responsibility.

Amanda hung out with Charlie, and their friendship was getting very close. Charlie could imagine asking Amanda to the movies again, and thinking this time she would say yes.

Pete was the one left out. He stayed in his room working on projects his father came up with. Pete's dad had him write a paper on the reason for being in the cemetery, and then another on how to be a responsible person. Pete wanted to please his dad, but he needed his dad to listen to his point of view as well.

When the leaves were almost off the trees and the wind blew the last ones away, Pete became anxious. He had to have his freedom back. He had to regroup with his friends. The only way he was going to be released was by the help of his grandma, but he couldn't risk asking for her help. His father might get angry with her, and pack up the family and move away.

One particularly chilly evening, when the wind was making a whistling sound against the house, Pete gathered the courage to speak to his dad. He crept down the back stairs and went into the living room. Pete's mom and sisters were playing a game, and Pete's dad was reading the paper. "Dad, can I talk to you?"

He lowered the paper and raised his eyebrow. "Not here, in the kitchen."

Pete's dad got up and followed him into the kitchen. Pete drew in a deep breath while looking for the courage to speak. "Dad, would it be okay if I hang out with my friends this weekend?"

Pete's dad scoffed and turned around. He was leaving the room when Pete started again, "I did everything you asked me to do. It's been almost a month, and I was wondering if you would let me see them." He dropped his voice on the last three words when he saw his father's expression.

"Pete, I don't plan on ever letting you see those kids again. What kind of kids hang around in a cemetery digging up a grave?"

Pete was crushed. He looked down to hide the tears welling up in his eyes. Large tears dropped to the floor. He turned to leave, and heard his name. It wasn't his father calling him, it was his grandma. "Pete, Peter, please come back here."

He turned, and she saw the tears streaming down his cheeks. She collected him into her arms and turned to talk to Pete's father. "Now you listen here, this is a good boy and he has good friends. I knew he was out there in that cemetery that night. I was the one who told him that there was no body in that grave." She nodded yes, "Yes, it was me." She patted Pete on the back.

"What kind of crazy mess have you both gotten into?"

Pete pulled away from his grandma, "It's not a mess, it's a mystery, and I want to solve it."

"It's the treasure," said grandma. "Pete found out about the treasure and he and his friends are looking for it."

"I figured that much when I was in the cemetery. Pete, when I was a boy, I too dreamed of finding the treasure. My friends and I would spend hours in the church and cemetery looking for the treasure too. We never found a thing, and neither will you."

"That's not true," said grandma, "They're already half way there."

"How do you know," asked Pete's dad.

"Because they found my daddy's Bible. I feel it in my bones, Pete is going to find it, and I am going to help him."

"So you think I should let him hang out with those kids? Do you know they dug up that grave?"

"No dad, we didn't. It was like that when we got there," insisted Pete.

"That means someone is one step ahead of you, and it could get dangerous," argued Pete's dad.

"No, we have a clue no one else does." He was immediately sorry the moment he said it.

"Oh, and what would that be?" asked Pete's dad.

Pete choked and said, "We, we want to keep it a secret so that no one beats us to the treasure."

Pete's dad came closer and took off his glasses. He was ready to say something when grandma spoke up. "I think is should be a secret, as long as it doesn't hurt anyone or anything."

Pete smiled at his grandma, thankful for her help. "It's not hurtful to anyone or anything, I promise."

Pete's grandma turned to face his dad and said, "See, there is nothing to worry about." She smiled and took his dad's hand, "Now are you going to let him see his friends or not?"

"Sure, under one condition."

"What ever you want, I will do it," said Pete.

"You stay out of trouble, and tell me where you are going. I want to be able to prepare if I have to talk to the police."

Pete and his grandma laughed out loud.

"I'm not kidding. You tell me where you go, and you stay out of trouble."

Pete hugged his grandma, and then walked over and hugged his dad. He left the room and they could hear him run up the back stairs.

Pete's grandma hugged Pete's dad and said, "Now isn't it nice to see him happy?""Sure, it's great until we get the phone call from the police.""

Pete could hardly contain his excitement. He made plans to spend the day with Davy on Saturday. Davy had committed to do some chores at Mrs. Elders, and Pete volunteered to help. Afterwards, they had planned to meet at Charlie's and talk about the clues and what they should do next. Billy was a little put out that Pete was going to hang around with Davy. He and Davy had become good friends, and he was jealous of Pete.

Pete was ready to leave the house Saturday morning when he saw his grandma in the laundry room. Pete had the little, yellow paper in his pocket. Davy had let him hold on to it, and he felt it in his pocket as he was leaving the house. He wondered if grandma would know what this means. At least they would have more information when he met with the group later today.

He quietly went into the laundry room and stood motionless behind his grandma. When she turned around and saw him, she screamed. "Grandma, what is it? What's wrong?"

"Oh boy, you scared the living tar out of me. Why didn't you tell me you were there?"

Pete smiled, "Sorry Grandma, I wanted to ask you a question?"

"Sure, what is it?"

"Grandma, when we were in the cemetery, we found another clue inside the headstone."

"Is this the secret clue," she asked.

"Yeah, and we can't figure it out." He took the small, yellow paper out of his pocket and handed it to her. She opened it and read the word S H E. "Hmmmm," she said, and turned the paper over and looked at the letters on the back. "I think this refers to the position of the stones on analemma on the door."

"That's what we think too. See T for top, B bottom, R right and L left."

She nodded in agreement.

"We also think that the second M is for Morganite. We think there are more stones, and we don't know where they are."

She nodded again, "I think you are really onto something." She turned the paper over again and read the word S H E. "I wonder if this is really a word or just someone's initials."

"Yeah, but whose?"

"Let me see. Who had a last name that starts with an E?"

Pete's eyes got wide. "I know someone, Mrs. Elders. Do you know her first name?" asked Pete.

The corners of her mouth came up into a very wide smile. "I sure do. She's my cousin. I used to play with her as a child. She's a very nice woman."

"Grandma, what is her name? Does it start with an S?"

"Pete, her name is Shirley Henrietta Elders. She never took her husband's name. I often wondered why."

"Shirley, that's an S, Henrietta, that's an H and Elders. That's it grandma, she's the S H E." He hugged her quickly and ran out of the laundry room. He was at the back door, and then suddenly stopped. He ran back into the living room, and then the kitchen, and then he yelled, "DAD."

An answer came from upstairs. Pete ran up the back stairs and down the hall. His dad was putting on his shoes in his bedroom. Pete stopped at the door out of breath. "Dad, I'm going to help Davy do chores for Mrs. Elders, and then I'm going over to Charlie's house." He paused, "I can go, can't I?"

Pete's dad smiled, "Sure, get out of here."

Davy was already at Mrs. Elders' house. Today they were going to clean out the attic. Mrs. Elders was concerned that the old stuff in the attic could be a fire hazard. Davy set up the ladder, and had the attic panel removed. "Is there a light up here?"

"I think the light is on the main beam overhead. There should be a string hanging from it," she replied.

Davy cautiously held onto the ladder and hoisted himself onto the platform of the attic. Streams of light came in from the vents. Cobwebs hung from the ceiling. Davy reached up and pulled it, and the small room was filled with glaring yellow light. He squinted because it hurt his eyes. Mrs. Elders had given Davy a large bucket with a rope attached. Davy was to put everything he found into the bucket and lower it down to her. She would take it out and he would pull the bucket back up. If an item was too large to lower in the bucket, then Pete would carry it down the ladder when he came over. Mrs. Elders was going to pay them ten dollars each for their hard work.

Davy didn't mind, he like exploring basements, garages and attics. He stood up and put his hands on his hips. "What's it look like up there?" yelled Mrs. Elders.

"It's pretty dirty," and as he said it he pushed cobwebs out of his way.

He looked around and found some old photographs. He bundled them and put them in the bucket. He found some old table cloths and towels, and quickly picked them up and shoved them in the bucket. When he did, his eye caught a glimpse of a fancy S H E. He stopped and stared at the small, fancy towel. Why would she have a towel with those letters.

"Are you ready to send down a bucket full of stuff?" she asked.

Davy snapped out of his trance and replied, "Yeah, let me hoist this down to you."

He took the bucket, let it hang in the opening, and then slowly released the rope until the bucket was firmly in her hands. "What is that on that towel?" asked Davy.

"Oh, these are my old, fancy towels. These letters," and she pointed to the S H E, "These are my initials. Nice aren't they?" She took the towels and pictures out of the bucket and said, "Okay, you can take it back up. Send me down some more stuff, let's get this going."

Davy took the bucket and continued to put small items in and send them back down. He was stupefied at the similarities. What if she is S H E, he thought?

Suddenly they heard a hard pounding on the front door. "That must be Pete. I'll go let him in."

Davy could hardly contain his excitement. He couldn't wait to tell Pete what he just found.

Pete practically ran into the house. "I need to talk to Davy, is he here yet?"

"Oh yes. He's already up in the attic."

Pete nearly ran to the ladder and yelled, "Davy, are you up there?"

"Yeah, Pete come on up."

Pete climbed the ladder as fast as he could and maneuvered his body on to the platform off the ladder. "I know who S H E is?" said Pete.

"Me too," said Davy.

"How do you know, I only just found out?" said Pete.

Davy held up another fancy towel with the initials S H E. "It's Mrs. Elders, isn't it?"

Wide-eyed they looked at each other. This is what the English teacher would call serendipity. "I wonder what else she knows?" said Davy. "Let's finish the attic, and then we can ask her when she gives us a snack."

"Okay," was all Pete said, and he got to work hauling all the items in the attic down to her.

They worked hard for two hours. Both boys were covered in sweat, dirt and cobwebs. When they finished, the attic was completely cleaned out. Davy and Pete were secretly hoping to find another clue in the old junk they hauled out of the attic. Pete put the cover back on the attic, and folded the ladder. Davy put the ladder back where he found it in the basement. He spent a few extra minutes looking around the basement for a clue.

When all was put back, Mrs. Elders was waiting for them in the kitchen with fresh baked chocolate chip cookies and milk. "Boys, you want to come in here for a snack before you leave?"

Pete finished washing his hands and was waiting for Davy in the kitchen. Davy came up the stairs disappointed that he didn't find anything in the basement. "Let me wash, I'll be right back."

Davy left and Pete sat down at the table and took the biggest cookie he saw. Mrs. Elders made the best cookies he had ever eaten. Davy came in and slid into the chair next to Pete. He grabbed a cookie off the top and took a big bight. He closed his eyes as the rush of sweet cookie and chocolate exploded in his mouth. He opened his mouth to take another bite, and stopped.

Pete noticed that Davy suddenly stopped eating and was staring at the nicely framed picture on the wall. He turned to look at it also. He had seen this picture many times and it wasn't particularly nice. It was of a house with a bike in the front yard. He looked harder and he saw it. Above the house was a figure eight. It was hard to make out because the glowing objects were faded and barely visible. It was similar to the photo in the textbook at school, but this was a painting of the analemma.

Pete looked at Davy and Davy looked back at Pete. Mrs. Elders was watching the entire process. "So you like my painting. I painted that when I was twenty seven. That in the sky is called an analemma. Do you know what that is?"

Davy nodded yes, "It's the sun at noon throughout the year."

Mrs. Elders looked at him in shock, "You are a very smart boy. Did you learn that in school? What they teach these days. It is just amazing."

"Can I take a closer look at the painting?" asked Davy.

"Sure, of course you can."

Both boys jumped up and ran to the painting. "It was right here in front of us all along," whispered Davy.

Pete studied the analemma in the sky while Davy combed the rest of the picture for the small 8. When they didn't find another 8 in the picture, they looked at the very ornate frame.

"Isn't that frame beautiful," she said as she saw the boys inspecting it. "My daddy made me that frame. He said it was special, just for my painting. In fact," and she gave a shy giggle, "It was my daddy that suggested I paint the analemma."

Somehow this new information was very exciting. Davy took the left side and Pete the right side. Together they combed the frame, and then Davy said loudly, "I found it!"

He put his finger on the edge in the lower left corner. "It's right here."

"Davy my boy, just what have you found?"

Davy realized that he and Pete must look rather impulsive to jump up and suddenly inspect her painting. He turned to face her and said, "Mrs. Elders I think this is the clue we have been looking for." How would she react? Would she be confused, mad, excited? He wasn't sure, but he had to say something.

She showed very little, in fact, no emotion. She got up and came over to where he was. She put on her glasses and looked down to where his finger was held. "Well I'll be. It sure is."

Davy and Pete just looked at her. She took her glasses off and sat down at the table. "How did you know where to look?" she said.

Taken off guard, Davy said, "Do you know about the 8's?"

"Why of course I do. All of us do. Your Grandmother, Charlie's Grandmother, and half the town who is related to us. We all know. What led you to me?" she asked.

Pete took the small, yellow paper out of his pocket and unfolded it. He slowly handed it to her. She took the paper and saw the S H E written on one side. "That's me?" she asked. She looked up at the boys. They nodded yes. She turned the paper over, and saw the letters written on the back. "I take it you know what this means?"

They exchanged a look, and Pete replied, "Not all of it. We think we know what some of it means, but we're missing something. We were hoping to find it here."

"And I think the painting is what we've been looking for," said Davy.

"Can we take it off the wall and see it better?" asked Pete.

She looked at the boys. She was getting caught up in the excitement. "We better, or we will never know the secret."

She walked over to the wall and started to take the painting down. Pete stepped in, "It may be heavy, let me help." He put the painting on the table, and they all looked at it for a minute. Then Pete turned it over on the back and looked. It appeared as if there was a small notch in the same corner as the 8. Pete ran his finger over the notch, and a small panel snapped backwards. They all jumped!

"Wow," said Davy, "that was cool." He put his fingers in the opening and felt material. Gently he pulled out a small dark sack made of velvet. It was very soft to the touch. He put it on the table and put his hand back in. There was another small sack. He carefully removed it, and then put his hand back in. "There should be three," he said.

"Why three?" asked Mrs. Elders.

"Because, we already have one," said Pete.

She looked at both boys surprised. "Why am I not shocked? You are the smartest, most resourceful boys I know."

Davy pulled out three sacks, and then just for curiosity, put his hand back in to see if there was anything else. It was empty. He pushed the compartment closed. Pete picked up the first sack and untied the draw-string. He carefully poured the contents into his hand. It was a brilliant, blood red stone faceted the same as the Morganite. "This is it," he said in a trembling voice.

Davy had the second sack open, and poured another stone into his hand. This one was green on the outside and red in the middle. He held it up to the light and said, "It's magnificent."

Mrs. Elders had grabbed the last sack and poured the last stone into her hand. This stone was half purple and half yellow. She held it up to the light and said, "How beautiful. Did you boys know these stones were here?"

Davy was holding up the green, red stone when he replied, "We have been looking for them for a while. I just found out this morning they might be here. It was the initials that gave you away. I found them on the towel, remember."

Mrs. Elders looked at the paper on the table.

"Yeah, and I found out before I left my house today. I asked my grandma if she knew anyone with those initials, and she said it could be you."

All three exchanged the stones and examined each one. They sat at the table for a very long time. No one said a word.

"Okay, so what do you do now?" she asked.

Davy looked up at her, "Mrs. Elders, we need to borrow these stones if we're going to find the treasure. We'll get them back to you, I promise." His eyes were pleading with her.

"Well, if you think it will lead to the treasure," and she smiled at him. "What do I have to lose? It's not like I knew they were here, so I'm not going to miss them. You promise you'll bring them back?"

"Oh, we promise," interjected Pete sitting up very straight. "We promise."

Pete and Davy put the stones back into the soft, velvet sacks and stuffed them in their pockets. Pete had one, and the small piece of paper, and Davy had the other two. They ate their cookies and ran out the door. In all of the excitement, she forgot to pay them, but neither one ran back for payment. They ran as fast as they could over to Charlie's house to share this information with the whole group.

Billy and Amanda were already there. When Pete and Davy came in, they motioned to the basement so they could talk in private. They crept quietly down the steps and went to the back, in a quiet corner. Pete and Davy looked around to make sure no one was watching, and then they emptied their pockets on the floor. The three velvet sacks and the small, yellow paper lay in the center of their small circle.

"What is this?" asked Charlie.

"The rest of the stones," replied Pete.

Both Pete and Davy were already loosening the ties to the velvet sacks and pouring the contents in their hands. Charlie grabbed the third sack and did the same.

"Where did you find these?" asked Amanda.

"You're not going to believe this," said Pete. "Both of us found out, entirely on our own, that S H E is Shirley Henrietta Elders, and the stones were hidden in her house."

"Yeah, she didn't even know about them," said Davy. He held the green stone up to show the red in the middle. Pete held the purple stone up to show that half was purple and half was yellow. Charlie held up the blood-red stone.

"Amazing," said Billy. That was the first word he said since they all met in the basement. "Do you know what this means?"

"Yeah, we're going to be ready," said Pete, and looked considerately at Billy.

Billy's eyes met Pete's, and they held the gaze momentarily. It wasn't exactly friendly, but it wasn't hateful either.

They exchanged the stones and held them up, and chatted about what they thought was going to happen when they put the stones in the analemma on the door of the church on December 25. The door to the basement closed softly, as not to let the five in the basement know that someone overheard their remarkable find.

Chapter Twenty - Nine

Old News

He walked around the inside of the musty church. "Uhhh," he said as his whole body shivered. The air had turned cold and temperature the inside of the church was chilly. He could see his breath if he exhaled hard. He stopped and stared at the stained-glass windows. Streams of light were coming through making shadows of blue, green and red on the pews. He jumped when he heard the door open.

The older man in the long, black coat entered through the secret door behind the alter. Mike Sullivan turned quickly to see him come in. "I didn't know there was a door there. I've been climbing through the open window."

The older man gave a grunt. He stopped right in front of Mike and glared into his face. "So what do you have for me?"

Mike dug his hands deep into his empty pockets, "Well, you see, I interrogated those monster kids and found out what was in the package they found in the church."

He was ready to tell about the Bible when he was interrupted, "It was a Bible, I already know. What else did you find?"

Mike shrugged his shoulders, "We did what you told us to do. We dug up that grave. You know the one that said Anita Fairchild. Are you sure you gave us the right grave because there was nothing there. Not even a body." He gave a little giggle, "Those stupid kids got there right after we left, and I watched them. One of them got in trouble. His dad showed up and yelled, pitched a fit, and drug that dumb kid off. I was laughing the whole time." He continued to chuckle to himself.

"Did you cover the dirt back over the grave and leave it as you found it?"

"UH, not yet. You see, we ran out of time and figured no one comes up here except those stupid kids."

"You fool, I told you to make it look as if you were never here." As he said it, he swatted at the young man's head. He pulled the young man's collar and got very close to his face. "You get your shovel and you fix that grave. You do it today. Do you understand me?"

Mike's eyes were wide and scared. He put his fingers over the man's hand and tried to pry them off his collar. "Settle down old man," and he pulled them away. He brushed his shirt down and fixed his jacket. He rolled his shoulders and stepped back from the man's searing gaze. "Okay, okay, I'll do it today."

The older man put his hands in his pockets and said, "So really, you have nothing for me. No Bible, nothing in the grave. What about that piece of paper the kids had?"

Mike sighed, "Nope, I tried, but they wouldn't give it to me."

"So basically you're ineffective. You didn't find anything." He stepped in again so that he was in Mike Sullivan's personal space. "Listen here you worthless piece of crap, those kids have again outsmarted you. They have all the stones and the date to use them. What I need now is for you to be here Christmas day at twelve noon, sharp. Bring a couple of your friends, hide in the church and wait for them to get here. Can you do that you meat-head?"

Mike coward under the older man's glare, "Yeah, sure, no problem. We can be here at 12:00 on Christmas day. Why not the day after or before, what's the big deal about Christmas day?"

"Because that is the day that solar noon and clock noon are the same. That is the day," and he pointed to the back of the church at the analemma on the door, "indicated by the figure-8. Now are you going to do something right or not?" He was raising his voice.

"Yeah, yeah, I'll be here and I'll bring my friends. It might be a little tough to get them to come on Christmas, but someone will be here with me." He walked away, and then turned and said, "Are you sure we shouldn't be in here at 11:00 a.m., you know, for daylights savings time and all."

The man grunted again, "You are the stupidest person on this earth, aren't you? Daylight savings time is in the summer time, not the winter. You need to be here at 12:00 noon, AND NO LATER!" He shook his head and stared at the floor with regret. "This is the last time I ask for your help." He turned and headed for the small hidden door.

"Hey, what about my payment? You said that there would be a lot of money if I helped," yelled Mike after the older man.

He turned abruptly and replied, "You said the right word, HELP. When you actually help, then I will pay you. Now get out there and fix that grave!" He opened the door and disappeared.

Mike gave another shiver and muttered to himself, "That old man creeps me out. I'm sorry I said I would help." He took in a deep breath. "If he weren't my uncle, I'd hurt him." He walked over to the hidden door and looked for a door handle. The area was smooth, no handle. He pushed on the panel, it wouldn't move. He pushed harder, nothing. He ran his fingers over both sides of the panel, again, nothing. He turned and headed for the window.

CHAPTER THIRTY

TOURMALINE, GARNET & AMETRINE

"So which one goes where?" asked Davy. He was in Pete's bedroom and they were sitting on the floor looking at the stones and arranging them according to the diagram.

"What does the T stand for?" asked Pete.

They had a mineral identification book out in front of them. "Look up the minerals that begin with a T in the index, and let's go through them."

Davy turned to the index and began to read the words out loud, "Talc, tantalum, tephroite, thorium, thulite, tin, titanium, topaz, torbernite, tourmaline, tridymite, tuff. These words are hard to say."

"We need to start looking them up. Look up the topaz. I see that in all the ads from the jewelry stores," said Pete.

Davy thumbed through the book and found topaz. "It says it has a hardness of 8, and can be colorless, white, yellow, blue, green, red, pink, violet, and brown. It has a vitreous luster."

"What does that mean?" asked Pete.

Davy thumbed through and found vitreous and read, "glassy, looks like glass." He picked up one of the stones, and then another and said, "They all look glassy. This is going to take for ever."

Pete reached over and grabbed the book out of Davy's hands, "Sorry buddy, but you take too long."

Davy's smile faded, and he chased one of the stones around the carpet. "I think we should show these to your grandma. She knew what the first one was, maybe she can tell us what the rest of them are too."

Pete looked over the book at him. "You know, I think you're right." He got up off the floor and opened the door, "Cover them up with the towel until I come back." He returned shortly with his grandma right behind him. "Grandma, these are the new stones I told you about. Please look at them and tell us if you know what they are."

Davy pulled the towel away and revealed three new stones, all the same size at the Morganite. Grandma got down on her knees and picked up the first stone. She held it up to the light. It was green on the outside and red on the inside. It reminded Davy of a piece for watermelon.

"Doesn't that one look like a watermelon?" asked Davy.

"Yes, yes it does. In fact Davy you got it right on the head. If I'm not mistaken," she picked up the mineral book and flipped through it. She appeared to be reading, and then she turned to another page and stopped and read some more. "Yep, I was right." She turned the book so both boys could see the page and showed them a tubular crystal that had red on the inside and green on the outside.

Davy got very close and read the name, "Tourmaline, wow that's pretty."

"And is often called, Watermelon Tourmaline because of the striking color. This one is very large. It's probably worth more because of its considerable size and clarity. Where did you find it?"

Davy gave a big grin, "In Mrs. Elders' kitchen."

"You mean to tell me that she had these stones, all three in her kitchen all along?"

"Yeah, and she didn't even know it," Pete chimed in.

"They were hidden in a secret compartment in a frame around a painting she did. She said her daddy made it for her special."

She held the stone up to the light, "Well I'll be." She picked up the blood-red stone and held it up to the light. "If I'm not mistaken, this one is a Garnet. If it were a ruby, it would be a true red, this one is more purple." She looked up garnet in the book and showed the boys.

"That makes sense because one of the initials has a G," said Pete.

"What initial are we missing?" she asked.

Pete shuffled through the objects on the floor looking for the small, yellow piece of paper. "Where is that paper?"

Davy reached into his pocket and held it up, "Here it is." The paper had become more worn and wrinkled in the last week. He read through the letters. "M for Morganite, but we knew that. T for tourma-something."

"Tourmaline," said Grandma.

"Yeah, that," he said. "G for Garnet, and A is for what?" asked Davy.

Grandma picked up the last stone. It was almost completely transparent. One side was light purple and it faded into light yellow. It sparkled when she held it up to the light. She picked up the book and turned to the index. Then she turned to another page and read. "Just what I thought."

"What is it Grandma? Do you know?" asked Pete.

"Well of course I know, I can use a mineral guide. This is not too unusual. It is called an Ametrine, this means it is part Amethyst and part Citrine. The only thing that makes this mineral change color is an impurity in the mineral make-up. They are very pretty stones." Again she held it up to the light.

"Well, mystery solved. When are you going to put the crystals in the analemma?" she asked.

The boys averted their eyes. Davy picked up the mineral book and started thumbing through it, and Pete started putting the stones away. She awkwardly got up off the floor and wiped her hands on her apron. She started towards the door. She put her hand on the door knob and started to turn.

"Wait," said Pete. He looked at Davy and said, "I have to tell her. She has helped us so much, she deserves to know." He turned and looked at his Grandma. "Grandma, you can't tell anyone, I mean anyone."

"I know Pete, I won't. You know that." She turned and sat down on the bed.

"Grandma, the next time we go to the church with the stones is on Christmas day."

"Isn't that going to be hard to do?" she asked.

"Why would that be hard to do?" asked Davy.

"Well, don't you usually spend the day with your family on Christmas?" she asked.

Davy look confused. For the first time he considered that it might be hard to get away on that date.

Pete looked at his grandma, "Do you think I will be able to leave and go to the church Christmas day?"

"I don't know," she replied.

"Pete, we will have to think of something," said Davy.

Chapter Thirty - One

Game Set

They had the stones, Tourmaline, Morganite, Ametrine, and Garnet. They had the location of each stone on the analemma. They had the date, December 25th. The only problem was making it to the church that day. Everyone agreed that they needed a special event to get out of the house at 12:00 noon and get to the church on Christmas day.

Christmas is the holiday that kids can't wait for it to finally arrive, and parents have too little time to prepare. Time was flying by, and they needed a good alibi.

"How about a Christmas pageant?' asked Amanda. "My mother would let me go to that."

"The problem is, she would probably want to go with you," replied Charlie.

"Good point," said Amanda.

"How about Christmas caroling? My mom would let me go do that," said Billy.

They all stopped and looked at him, "I seriously doubt your mom would believe you were out caroling," said Pete. Everyone gave a laugh, even Billy. "Yeah, you're right, she wouldn't believe that."

"It has to be something that's admirable, yet believable," said Charlie.

"Yeah, and something that no one else wants to do with you," said Amanda.

A low grumbling sound came from Davy's stomach. He grabbed at it and said, "Be quiet in there, no one can think with all that noise." He looked up at the group and said, "I really could use a sandwich."

"That's it," exclaimed Pete. "We could make and hand out sandwiches to the homeless on Christmas day. A friend of mine and her whole family does it every Sunday. Except on Christmas we start early, and end at about 11:30, just in time to be in the church before the noon."

"My mom would let me do that," said Davy.

"I think they would believe me doing that. The only thing is that I would probably eat half of them before I gave them away," laughed Billy.

"What about you Amanda?" asked Pete.

She nodded her head, "It might work."

Charlie was the only one who did not respond. He was staring off into space.

"Charlie, you didn't answer. Do you think you could use that excuse?" asked Amanda.

He shook his head no, "I'm afraid I may miss the whole thing."

"Why?" asked Pete.

"My mom is planning on taking us all skiing in Denver over Christmas. The funny thing is, my stepdad is staying home to take care of the business."

"Is there anyway you can stay home too?" asked Pete.

"I'm trying to find a way. They told us last night, and ever since I have been racking my brain to come up with a solution."

"I guess handing out sandwiches is not enough," said Davy. Billy laughed out loud at his comment, and then saw Charlie's face and immediately stopped. "Sorry buddy," he said.

"The only way I would be able to stay home was if someone was ill, or dying," said Charlie.

"Don't say that, it's bad luck," said Amanda.

"I didn't mean..." and he shook his head while he tried to finish his sentence.

"You know, that's not a bad idea," said Pete.

"Yes it is, no one wants anyone to get hurt, be sick or die," retorted Amanda.

"Not anyone, Charlie." Pete took a deep breath and laid out the plan. "Charlie, you need a fever and a rash. Rashes are always bad and most of the time they won't let you on the plane with a rash. Add the shivers, sweats, and maybe some diarrhea and you're staying home."

Charlie's eyes widened. "Pete, I think you're right. I may be coming down with something, and I'm quite sure it's contagious. Too bad my illness won't begin until we are ready to leave, and end as soon as they're gone." He chuckled. "I'm going to have to act like I want to go real bad. I don't want to give away the plan."

"Yeah, and you're going to have to be very disappointed you can't go," said Billy.

"Tears really work," said Pete. Everyone stopped and stared at him, "What?" he said as he looked back at them.

"Okay then, we set the stage. We all go home and start laying the ground work for Christmas day, and we need to make it believable," said Pete.

"We will need to buy bread and lunch meat and make real sandwiches. We need to pack them up, add some chips, cookies and sodas, and we will be out the door," said Davy.

"Okay, everyone agree. Go home, lay the ground work. Make it believable, and say we will do it as a group," said Pete.

"Yeah, my mom would be much happier to know I am going with a group of kids instead of by myself," said Amanda.

Pete looked at her, suddenly realizing they really were going to do something nice for others before they went to the church. It hit him that the plan was really about more than just them, and that serving others should be part of the whole plan, not just the alibi.

"Let's make them really good sandwiches, and good cookies too," he said.

"Who cares what kind of sandwiches we make?" asked Billy.

"I do, because someone will want a nice sandwich on Christmas day," replied Pete.

It took some explaining, but Pete managed to convince his parents to let him go to the local homeless shelter and hand out sandwiches on Christmas day. Pete's mom thought it was a wonderful idea, and she was very proud of Pete for volunteering to do it. Pete's dad also seemed impressed, and actually smiled at him during dinner.

Davy's mom was concerned as to the kind of people he might come into contact with. Davy convinced her that they were normal people, just down on their luck. Davy's dad was pleased that Davy was thinking about others during the holiday season. "So many kids think only of themselves and not of others, I'm very proud of you," he said. Davy felt guilty considering they were doing this good deed to cover for their selfish interest. He gave his dad a weak smile, and nodded in agreement.

Amanda's mom almost said no, but reconsidered when Amanda told her that she would be with a group of kids and they all got approval from their parents. She agreed, but wanted to drive them. Amanda insisted that they already had transportation, and she didn't want to take her mom away from the Christmas celebration with the family. "Really mom, I'll be just fine. I'll carry your cell phone with me, and call you if I need you." This pleased her mother, and she was good to go.

Of all the kids, Billy's mother was the most impressed. She hugged him and cried. "This is the best Christmas present ever," she said as she tightly squeezed her son. "Mom, it's not that big of a deal. All I'm going to do is hand out sandwiches to less fortunate people Christmas day." After he said it, he realized what a nice thing is really was that he was going to do. He realized that what Pete said was true. Someone is going to eat these sandwiches and be happy to have them.

CHAPTER THIRTY - TWO

A BAD RASH

Charlie was the only one who didn't have to come up with a reason to leave the house on Christmas. He patiently listened to the others tell of how they were all set to go. "My mom bought buns, muffins and packages of cookies. She's going to help me make the sandwiches and get ready. I'm not sure I'm going to be able to carry everything she bought," said Billy. He was really getting into the charity part of the plan.

Amanda looked at Charlie's face. He looked sad and almost worried. "Here, I brought this for you. My mom gave me her old make-up, and I thought you might need it, you know, for the rash." She held out a small, flowery make-up bag.

Charlie looked at it with repulsion. He even backed up a little as to not have it near him.

"Charlie, take it. I brought it for you to use," insisted Amanda.

The other boys stared at it with curiosity. No one said a word. It was as if she was holding out the most disgusting spider, and no one could believe it. "It won't hurt you, just take it."

"I don't think he wants it," said Pete coming to his rescue.

"It's not like I'm asking you to wear it to school, I'm just trying to help you find the perfect idea for your rash. Don't you need a good rash to pull this off?" She pulled the small bag back, opened the zipper and pulled out three tubes of lipstick. The first one was bright pink with sparkles. The boys watched as she twisted the bottom and the lipstick shot out the top. "What do you think about this one?" She swiped her finger over the top and then rubbed her fingers together. "Isn't that pretty?" she said.

"Yeah, but what about the sparkles? Charlie's mom is going to think he's breaking out with a bad case of glamour," said Billy. They all laughed except Amanda. She gave a huff and then twisted the bottom and it disappeared inside the tube. She picked up the second one, twisted the bottom and out popped an orange-brown color. She swiped her finger and rubbed them together like she did before. "What about this one, no sparkles?"

"It looks like your fingers have a bad tan," said Pete. Again the boys broke out in laughter.

"I'm glad you all find this so funny," she complained. She put the tube away and picked up the third one. Charlie touched her arm, "Really, you don't have to do this. I can find a way to have a bad rash myself at home." They laughed again.

The third tube was deep red. The boys stopped and stared at the color of the third lipstick. She grabbed Davy's elbow, and ran the lipstick down the inside of his arm. "Hey, what are you doing?" he said as he tried to get his arm back. "Just a minute she said." She put the tube down and started rubbing the lipstick into his arm. It looked smeared at first, and then she took out some pressed powder and gave a few puffs over the lipstick, and presto, instant rash. They all leaned their heads in to get a better look. "I have to admit, it looks like a really good rash," said Billy.

Davy pulled his arm back with disgust, "Why did you have to do it to me? Everybody picks on the little guy." He began rubbing the odd mixture off of his arm. It just got redder and more spread out.

"I have to admit, that is a really good rash," said Charlie. He put out his hand and Amanda zipped up the make-up bag and handed it to him. "Do I have to take the whole bag? Can't I just take the lipstick and that other stuff?"

"Oh, just take the whole bag and make her happy," Billy practically giggled.

Charlie glared at Billy. It was not what he wanted to do, carry around a make-up bag. What if he got run over by a car on his way home and they opened his back pack, and there was an old make-up bag. He would have to die in the accident because he couldn't live with the shame. He put the bag in his back pack, shook it so it would fall to the bottom, and then zipped it up. "I need to forget it's in there," he said to the other boys. Pete nodded in agreement, he was glad it wasn't in his back pack.

School ended for the holidays, and they were set. They made a phone chain, and agreed on where to meet with their sandwiches. They also went over who was going to bring the necessary items to the church. They made it clear that if there were any problems at all, they had to start the phone chain.

Charlie's symptoms started to develop slowly. The day before they were to leave for their trip, he complained of a headache and a small rash started to develop on his arms. That night, he had the sweats and the rash had spread. It was all over his body, and he felt sick to his stomach. The next morning, he could barely get out of bed.

"Charlie, you can sleep on the plane and I'll take you to the Medstop in Denver," said his mom.

"I don't want to be a burden, I'll just stay home with dad," moaned Charlie in the most pathetic voice he could muster.

"Don't be ridiculous, it's probably just a virus. You'll be better in a day or so. I don't want you to miss your chance to ski."

Charlie was going to have to give the performance of his life. First, he needed to increase the rash, add some heat to his head, and then add a layer of sweat. He snuck into the bathroom and prepared. When his mom came in his room, he was ready. He was lying in bed and ready for his cue.

"Charlie you really need to get up and pack," said his mom. She was folding sweaters and putting them on a stack on his desk. Slowly he sat up. Grabbed his head and moaned. With extreme effort, he swung his feet to the floor and stood up. With one had on his head, he reached for the chair. He stumbled over to it, and then with a perfect launch from the chair, he began his fall. He let his legs buckle under his body while his arms did a half pinwheel by his sides. His performance ended with him sprawled out on the floor. He gave a weak moan as an encore.

After that, he was lucky he didn't end up in the hospital. His mother screamed, called for his stepdad. Together they put him back in bed. There was a phone call to the doctor, medication, soup, and of course he was staying home. His stepdad did not seem at all happy that he was staying home, but agreed to take care of him. Charlie was in.

On Christmas Eve, the final preparations were being made at Pete's house. His mother and sisters finished decorating the tree. Since they had a real tree, Grandma insisted they wait until Christmas Eve to put it up so it would last longer. Grandma helped them all string popcorn, and decorate real wreaths. The house smelled like Christmas. A few days before, they spent the whole day making Christmas cookies, and now a plate of overly decorated cookies, and a tall glass of milk was being placed by the fire place where the stockings were hanging. It looked like a scene out of the movie *It's A Wonderful Life.*

Pete was happy just having his dad home and everyone together. His anxiety was starting to rise as he anticipated the events of the next day. He was interested in Christmas presents, but he had to be very careful not be too excited. He did not want to draw any attention to himself or have his parents wonder if something was up

He went into the kitchen and started to take out the items he needed to make sandwiches. "Are you going to make the sandwiches tonight?" asked his dad as he entered the room.

"Yeah, I didn't want to take any time away from the celebration tomorrow," replied Pete.

Pete's dad rolled up his sleeves and washed his hands, "Then let me help you,"

Pete felt a lump develop in his throat, "Thanks," was all he could say. His eyes got hot and he felt overjoyed to have his dad help him. He was happy to have his dad take interest in him without judging, or giving his opinion.

Pete and his dad were busy making sandwiches while his mom, grandma and sisters got ready for bed. A big tradition with Pete's mom was to sing a few Christmas carols, and then everyone go to bed early. This gave his parents plenty of time to wrap and place the presents under the Christmas tree. When Pete was done making sandwiches, he wiped up the mess, put everything away, and got ready for bed. Before he got into bed, he got out the sun clock and checked to see if the Morganite was still there. He let out a sigh of relief to see it safely wrapped in tissue paper in the compartment in the back. He slid the sun clock under his bed and closed his eyes. He was too excited to sleep. He kept thinking about what happened the last time they put the stone in the analemma. He was wondering if this time they really would find the hidden treasure.

As his mind raced backed to the church, he drifted into a deep sleep. He eyes started to move rapidly as he started to have a crazy dream. *He was walking down the trail to the cemetery. Charlie was ahead, and the rest of the group was behind him. He felt snow crunch under his feet, and felt a cold wind in his face. They entered the cemetery and something was missing. The angel that held the magnifying glass was gone. He looked around in panic, and the gate with the sun emblem with the stone for one eye was gone also. How could this be? Suddenly, all the headstones were gone, and the cemetery was full of muddy holes. All the graves were disturbed. He ran to the window of the church that was usually open, and it was closed shut. He couldn't force it open. Pete put his face close to the window and he could see people moving around inside. A man in black was waving his arms and giving orders. He did not recognize the people in the church but could make out Mike Sullivan, who had his hands in his pockets and seemed to be crying. He banged on the window, but no one turned to look at him.*

He looked back at his friends, and they were gone. He turned to bang on the window, and it was now open. The people inside were looking at him. He backed away and fell over a rock. They came over to the window and Mike started to crawl out. Then the whole building started to move. The walls began to shake and the shingles were falling off the roof. Pete stood up and ran down the path. He looked back and the man in black was right behind him. He let out a scream and... woke up. Sweat was rolling down his face as his heart was racing wildly.

He got up and went into the bathroom. He splashed water on his face. He was so disturbed by the dream that he decided not to go back to bed. Instead he went downstairs and found the presents nicely stacked under the tree. The room felt so comfortable, that he lay down on the couch and put his head on the pillow. He closed his eyes and fell into a dreamless sleep. The sounds of squeals and laughter woke him up. His little sisters were standing in front of the Christmas tree touching all of the presents. Pete sat up and rubbed his eyes.

The morning was a wonderful event filled with nice presents, fine food and laughter. Pete was having such a wonderful time, but pangs of anxiety tore at him, so he went upstairs to get ready to leave by 10:30 a.m.

Meanwhile, Charlie waited for his stepdad to leave the house for Christmas Services. He got dressed, put pillows in his bed to look like he was still there, and then snuck out. He didn't have sandwiches or cookies, but he planned on meeting with the others anyway. He too left promptly at 10:30 a.m.

Davy and Billy walked together, and met Amanda before reaching their destination point. Charlie and Pete were waiting when the others arrived. "You don't look so good," Billy said to Charlie.

"This stuff doesn't want to come off," replied Charlie.

"Tell me about it," Davy chimed in. "It took me three days to get that stuff off my arm, and I scrubbed and scrubbed."

Amanda let out a giggle and let it end in a big smile. She looked around, "Shall we go. We don't have that much time before we have to be at the church."

They made their way down to the Goodwill Center, and opened the door. The room was filled with the smell of ham and roast beef. There was a long line of people waiting for a plate of food.

"They already have plenty of food, what are we going to do with all of these sandwiches?" complained Billy. He was truly disappointed that his sandwiches might not be needed.

"Can I help you kids?" came a pleasant voice from behind them. A nice elderly woman with gray hair and glasses was standing there.

"Um, we um, came to hand out sandwiches today," said Billy. "We wanted to help the unfortunate today."

"What a wonderful and generous idea," she said. "I am sure many of the people would be happy to have a sandwich to take with them today as they venture out in the cold. Here, let me help you."

She moved into the center of the hall and cleared her throat. "Today, on Christmas, we have some generous young kids who have come by with some extra food, I believe sandwiches, which they made and would like to share with you. Would anyone be interested in taking one?"

Heads began to turn and look at the kids standing shyly in the door way. An elderly man got up from the table where he was eating, and came over and reached out his hand. Amanda quickly handed him a sandwich. Other hands reached out from the tables where they were sitting. Davy eagerly walked over and started handing out his sandwiches. Billy followed doing the same. Amanda gave Charlie some of her sandwiches, and they moved to the other side of the room. Pete walked to the row of tables in the middle, and gave out his sandwiches. All the sandwiches, cookies, and fruit were gone in ten minutes. The kids gathered back by the door, and were ready to leave before 11:00am. The elderly woman was standing by the door and thanked them all personally before they left.

Pete walked back out into the cold followed by the rest of the kids. The sun was bright in his eyes. He had to squint as he looked around at the others. The reality of the cold hit him like the reality of the large number of poor sitting in the hall. "That was sad," he said.

"Yeah, I can't believe how many needy people there are in this small town," said Davy.

"And this is a small town, can you imagine what it must be like in a big city?" said Amanda.

Billy didn't speak. He just stood there. The look on his face was neither happy nor sad. "You okay?" asked Charlie.

Billy nodded.

"I know we did something really good in there, but I don't feel like I even made a difference," said Amanda.

Billy said in a rather sad voice, "Yeah, me too."

They stood there in silence for a few minutes, and then Pete looked down at his watch and said, "We better make our way over to the church. It's already after eleven."

They turned and moved quickly down the street towards the church.

Meanwhile, the church was filling up with other occupants. The first to arrive was a man in a long black coat. He entered through the secret door and hid behind the alter. The next to arrive was a small group of boys, three to be exact, through the open window. They clumsily dropped in, and discussed where they should hide. After five or so minutes of arguing, they agreed to hide in the pews in the middle of the church.

After a few minutes passed, the small, hidden door in the front of the church opened again, and an elderly woman in a cape holding a book entered. She swiftly moved to the second pew, and disappeared low into the seat. The church probably had more occupants that Christmas day than it had in many years.

The small group of kids entered the cemetery and were getting out the items needed to set the stage. Davy closed the gate and made sure it was securely wedged between the metal prongs. Billy fished the magnifying glass out of his pocket, and walked over to the angel. He put the magnifying glass into the slot and then glanced down at the grave next to the angel. The muddy hole that was once there was gone and the grave was again nice and smooth. "Hey, look at this," he yelled to the other kids.

They came walking over quickly, "What is it?" asked Davy.

"The grave, the one that was dug up, it's covered now," said Billy.

They stared at it for a moment, and then Pete said, "I think we should get inside."

Charlie nodded, "Yeah, we got to get things ready."

One by one they slipped through the window. The air inside the church was stale and chilly. All five assembled by the doors in the back. Davy sat down on the floor and took out the small, velvet bags. He handed one to Amanda and one to Billy. Pete took out the sun clock and opened the back. He took out the Morganite wrapped in tissue paper. Pete handed the small, yellow paper to Charlie, and they began to put the stones in the holes on the door.

"Here we go," he said. "I need the Morganite, Pete." He reached over and Pete held out the Morganite. He carefully secured it into the hole on the right. It fit perfectly. "Okay, I need the Garnet. Who has the Garnet?" he asked. Amanda's hand shot out, and she produced the blood-red stone. "Thanks," he said, and carefully placed it into the slot at the bottom. "Uh, now I need the A one. What is that stone?"

"Ametrine," said Davy and he handed it to Charlie. He carefully put it in the slot on the left.

"One more," said Charlie.

"It's Tourmaline," said Billy, "And it goes on the top."

Charlie looked at Billy, "No kidding, on the top." He took the stone from Billy with a big smile.

"At least I don't look like a sickly tomato. That's some rash you got." Billy retorted.

The rest of the kids laughed. "Ah, very funny," Charlie shot back.

"What time is it?" asked Davy.

Pete looked down at his watch. They had been moving rather slow and he was surprised to see that it was about ten minutes to 12:00. "Wow, it's almost noon," he told the other kids. "What else do we have to do?"

"I think we're ready," said Amanda.

"Do we need to have someone outside to make sure things stay set?" asked Davy.

"I don't want to be outside," said Pete. "I want to see what goes on in here."

"Me too," said Billy. "Last time I was outside and I missed all the excitement."

"I want to stay inside," said Amanda.

"I want to be inside too," added Davy.

"Well, if you think nothing is going to go wrong, then let's all stay inside," suggested Charlie.

"Wait, do we need the sun clock?" asked Amanda.

"We shouldn't" said Pete. "Clock noon and solar noon are the same on Christmas day."

"Is the sun out?" asked Amanda. "I can't remember if it's cloudy or sunny."

They looked at each other stunned, "What difference does that make?" asked Billy.

"We need the sun for the whole thing to work," said Pete. He walked quickly to the window and looked outside. Davy pointed to the streams of brightly colored light coming in through the stained glass windows. In unison they said, "Yeah, it's out."

Pete turned to join the others in the back when he spotted someone crouched down in a pew in the middle. He froze.

Charlie was watching Pete and said, "What is it?"

Pete only pointed to the pew. The person ducking down did not see him point. Quietly, Charlie and Billy moved down the middle isle to the same pew. They stood there looking at the person crouched down, hiding. He popped his head up and saw three boys staring down at him. Quickly he shot up and said, "What are you puke-breath morons doing here?"

"Well, we're not hiding in a pew," replied Billy.

Mike Sullivan was standing with his hands on his hips between them. "Shut up Billy, you fat butt, stupid jerk."

Billy stepped in front of Charlie and was ready to respond with his fists when two other heads popped up from the pew behind Mike and the other in front of him. "Yeah, go ahead, the odds are a little different today," Mike boasted. "We plan to take the treasure the moment it's revealed."

"You aren't taking anything," said Charlie.

Mike threw his head back and laughed, "And you're going to stop me?"

"All of us will stop you," said Amanda.

Mike turned to look at her, "That will be fun," and he laughed.

Pete glanced down at this watch. Only four minutes till noon. What were they going to do? He could run back to the door and take the stones out and they could try to leave. He looked up at Charlie, and saw Charlie staring at Mike. Mike was still looking at Amanda, and that was when Charlie lunged at him. He threw his body onto Mike's midsection, and he fell hard against the pew. The pew in front of Mike fell over into the boy standing there, and he fell over with it. Billy jumped on the kid in the back pew and knocked him to the ground. He climbed on top of him and sat with his knees over the boys shoulders. The boy could not move. Davy grabbed the legs of the boy in the pew in front of Mike, and he held him securely under the pew. The boy was flat on his back with the pew resting on his stomach.

Charlie pushed Mike Sullivan onto the floor, and pulled his jacket up over his face. Pete jumped in to assist Charlie. They had Mike pinned down, and he could not see what was going on.

"What should we do?" yelled Pete as he struggled to hold down Mike.

"What do you mean?" Charlie shot back.

"Do we want to take the stones and leave?" cried Pete.

"No, it's almost noon," yelled Billy. "I want to see the treasure."

"Me too," yelled Davy, as he struggled to keep the boy's feet still.

Amanda shifted from one foot to another, unsure of what she should do. "Do you want me to yell for help?" she asked.

"There will be no calling for help," Came a low voice from behind her.

She turned abruptly to see a man standing behind her in a long dark coat. "Let go of those boys," he said in a stern voice.

Charlie looked surprised. He knew that voice. He turned around to gaze at his stepdad's face. He let go of his grip on Mike Sullivan, and stood up quickly. "Did you come to help? How did you know we were here?"

As soon as Charlie released his grip, Mike rolled away from Pete. He stood up and adjusted his jacket. He swung around and punched Pete in the shoulder, and was ready to hit him again when another loud command came from the man. "Stop!"

Mike looked over at the man, "That's not what you said earlier, you told us to do what it takes to get the treasure."

Charlie stood there stunned. "You knew about this?" he said to his stepdad.

No one moved for a long moment. Then another voice rang out, "Of course he knew. He's been watching you and listening to your plans." They all turned around to see the little old woman in the cape standing in the second pew.

"Grandma," gasped Pete.

She nodded at him, "He's the one who told your dad you were in the cemetery the night the grave was uncovered. I overheard the conversation your dad had on the phone."

Charlie looked at his stepdad, "I have had this feeling that someone was watching me. Have you been spying on me?"

"How else would anyone know we were here? We didn't tell anyone," said Amanda.

"He must have heard us making plans in your basement the day we found the stones," said Davy. "That is the only way he would have known we were going to be here today."

The man shook his head, "And I will take the treasure out of here today," he said in a definitive voice.

At that, a white beam of light shot through the center of the church from the all-seeing eye behind the alter. "It's starting!" yelled Amanda.

The light hit the Garnet at the bottom, marking December 25. Red rings of light emanated from the Garnet. Another beam of white light hit the Ametrine, and purple and yellow rings followed. In the next second a white beam hit both the Tourmaline and the Morganite. Rings of every color flooded the church. It was amazing to see, but what followed next was nothing less than remarkable. When all of the rings of color met, a streak of light, similar to lightning, bolted to the reflective ceiling. From there the streak burst into smaller flares and fanned out. The churched was lit up like fireworks. The flares hit the pillars holding up the outside walls of the church. The church began to shake. Large hunks of rock and boulder began to fall. The flares hit the windows, and they began to rattle and break. Pieces of colored glass hit the floor and shattered into brightly colored shards. They could hear the shingles falling to the ground outside.

Everyone ducked. Pete threw his hands over his head and got down next to the nearest pew. He closed his eyes and pulled his head down farther. It suddenly occurred to him that he dreamt about the church shaking last night. It lasted for less that a minute, and then it stopped. Again, no one moved. Charlie was the first to stand up. "Is everyone okay?" he said. He saw Amanda look around, and he saw Billy stand up. He couldn't see Davy or Pete. "Pete, Davy, where are you guys?"

Clouds of dust rose form the debris. Davy pushed a large piece of glass off of his shoulders, "That was wicked," he said.

"Are you hurt? That large piece of glass landed on top of you?" asked Amanda.

"No, it wasn't sharp," replied Davy.

Pete stood up and brushed himself off, "Grandma? Grandma, are you okay?" he said as he made his way through the rubble towards her.

She was lying in the second pew. There were shards of broken glass and rubble on top of her. "Oh my gosh," yelled Pete.

Charlie and Billy saw her and were making their way over to where she lay motionless. Charlie's stepdad stood up and blocked them. He looked around at the church and said, "So, where's the treasure?"

Stunned, Charlie stopped and looked at him, "What are you talking about? She could be hurt," and he pointed to Pete's grandma.

Billy pushed passed the man and got over to her first. He started pushing rubble and glass off of her still body. Pete climbed over the pews to get to where she was. They carefully removed all the debris, and then

Pete put his hand gently on her head. He felt her neck, and bent down to feel her breath on his face. "Is she okay?" asked Billy.

"She's still breathing," replied Pete.

Watching this and hearing the news, Charlie's stepdad moved in closer to Charlie, "Now tell me where we find this treasure." His voice was loud and full of command.

"I don't know," replied Charlie. He looked around. All he could see was rubble, broken glass and pews. A huge cloud of dust made it hard to see across the room. It could be anywhere.

"Start looking, and don't stop until you find something," yelled the man in the black coat. "Everyone, NOW!" and he glared at Pete and Billy.

Mike Sullivan was brushing himself off. "Stop grooming yourself and look for the treasure!" the man yelled at Mike. "Okay, okay, I'm looking," he mumbled back.

All the kids, including the two with Mike started moving debris, and making their way around the church.

Amanda was scared, and stayed near the back where there didn't seem to be much debris. There was no secret door in the wall, no secret passage in the floor, nothing. Much more light came in since the windows were all broken. Even in the bright light of the noon sun, no treasure was revealed.

"Move these pews," yelled the man in black at the kids. The bigger kids picked up the ends of the pews and started moving them to one side. The thick cloud was beginning to settle. For almost an hour, they looked through the debris.

Pete, who did not leave his grandmother's side, spoke up. "I need to get my grandma some help. I think she might really be hurt."

"No one leaves until we find the treasure!" the man yelled.

"What if there is no treasure? What are you going to do?" asked Pete.

The man in the black coat glared at Pete. One of the other boys that came with Mike spoke up, "Hey, I don't think there's anything here. We looked through all of this mess, and there's nothing. I'm not sticking around," and he made his way over to the broken window and crawled out. The other boy looked around, and then followed him out. They could hear them walking away from the church talking about the flashes of light and rings of color.

Mike Sullivan dug his hands in his pockets, he wanted to leave too. "Hey old man, I'm going to take off." He scurried quickly to the window, and was about to leave, when he was called back.

"Get back over here. You are going to help me find this treasure," said the man in black. "All of you keep looking."

Charlie stopped searching and walked over to his stepdad. "So, if we find the treasure, you're just going to take it. What were you going to do after you took it? Were you going to just disappear?" Charlie was clearly distraught. The thought of betrayal was overwhelming. He didn't know what his mother would do if his stepdad just up and left.

To avoid Charlie's searing questions, he yelled, "Where is the damn treasure?"

"Can't you see, there is no treasure," Charlie yelled back with tears in his eyes. His breaths were coming in short gasps as he began to heave. "Why don't you just get out of here? Just leave us alone."

Amanda came along side of Charlie, and took his arm. He looked over at her embarrassed by the situation, and shook his arm away from hers. "You ruined it! Because of you, we will never find the treasure now! Just get out of here and leave us alone!" he shouted at his stepdad.

Davy stood up and walked over to Charlie. Billy came up from behind. All the kids stood in a semi-circle around the man. Mike stared from across the room. He looked at them, one at a time. He stepped back and looked at the church now a mess of rubble and broken glass. His face was becoming red and filled with anger. Amanda was afraid he was going to hit Charlie. He clenched his fists and yelled, "NO ONE LEAVES UNTIL I GET THAT TREASURE!"

Amanda backed up and the sudden movement made her the target. He reached over and grabbed her by the color. Mike Sullivan was suddenly overcome with regret and ran towards the man in black. Charlie lunged for him from the other side, and together they knocked the man off his feet. Amanda fell backwards into the pew, and Davy ran to help her up.

"I've had it," yelled Mike and he hit the man on the floor with his fist. "You are, without a doubt, the worst father, uncle, man, I have ever seen." And he hit him again in the face. Billy, stepped in and grabbed Mike's hand as he was going to hit him again. He shook his head no when Mike turned to look at him. "Leave him alone," he said.

Mike got to his feet and adjusted his jacket and ran his fingers through his hair. He gave a huff and said, "That's it, I'm out of here." He looked around at the group of kids, nodded his head slightly, walked quickly over to the window, and was gone.

The man in the black coat wasted no time getting off the floor. He smoothed down his coat, and looked around at the angry kids. He quickly strode over to the hidden door and disappeared. For a moment, the only sound was Charlie's sobbing. He wiped his face with his sleeve and turned so the others couldn't see his face.

Pete had been watching Charlie, and did not notice that his grandma was sitting up in the pew. "Poor thing," she said.

Pete turned at the sound of her voice, "Grandma, you're okay?" he exclaimed.

They all turned to see the woman sitting in the pew smiling.

CHAPTER THIRTY - THREE

FROM THE RUBBLE

"Do you think he's gone yet?" she asked. She got up and walked over to the secret door. She pushed a small lever, hidden from plain view, and the door opened. She watched as the man in the black coat left the cemetery, and then she turned around. Again a big smiled filled her face.

"Grandma, do you know where the treasure is hidden?" asked Pete.

"I sure do," she said. "It's right out in front of you."

With that, they immediately looked around. "I don't see anything," complained Billy.

Amanda shook her head, "Me neither."

Davy stood completely still. He was staring at the floor. Pete saw him and went over to where he was standing and looked at the floor. All he could see were shards of colored glass. "What is it? What are you looking at?"

Davy pointed to the floor, "Don't you see them? They're right in front of us." He walked over and picked up what looked like a small piece of broken glass. He came over and handed it to Pete. He moved the colored object to his face, and it was clear. It wasn't just pieces of broken glass, it was beautifully colored stones. Perfectly cut, and in all colors. There were emeralds, sapphires, rubies and diamonds. There were hundreds and hundreds of them. They worked for hours picking them up.

Caught up in the moment, they lost track of time. The bright noon sun turned into long shadows and it got harder to see. Pete looked down at his watch, "Oh my, it's almost 5:00. We need to get home."

"What about the rest of the stones?" asked Billy.

"Well, we don't tell anyone, and come back tomorrow. We can pick them all up if we work all day and move the pews," said Pete. He walked to the back of the church and removed the four stones from the analemma on the door. He handed three to Davy and said, "Don't forget to give these back to Mrs. Elders."

Davy nodded and said, "I think I'll share some of my stones with her too."

Grandma took Pete's hand and cleared her throat. "You all need to listen; I have something important to say." She looked around at all the kids. "My family fell apart and wasn't close because of this treasure. If you are to remain friends, you must trust each other and be honest about what you do." Her green eyes were radiant in the last rays of the sun. "This treasure can either bind your friendship, or end it forever. Be very careful how you handle it."

"I think she's right," said Amanda. "Let's take what we have here, and then come back and work together tomorrow. We can split up everything evenly when we have all the stones."

Heads nodded. The excitement of the day had taken the energy out of them. They left by the way of the secret door. On the way out, Pete turned to Charlie. "Why don't you come home with us tonight? Your mom is gone, and I really don't want you going back to that man, if he's still there."

Charlie nodded in agreement. "Yeah, okay, thanks."

They left the cemetery without another word.

CHAPTER THIRTY - FOUR

A PRECIOUS ENDING

Everyone was at the church early the next morning. The plan was to move every pew and sweep up every square inch of the church. Pete's grandma and dad came to help, so did Mrs. Elders, Amanda's mom, and Billy's mom. The stories of the earlier day's events were repeated over and over. No one complained, no one boasted, and every worked together.

When it appeared that all the precious stones were gathered, and no more could be found, the whole group left the church. Walking away, Pete looked back. He was sad to see the family church destroyed. Secretly, he wanted to restore it.

They assembled in Pete's kitchen and a plan to sort out the precious treasure was devised. They would get appraisals from at least three different jewelers to determine the worth of the stones. They would then be equally divided among the members of the group. Of course, they knew this was going to take some time, but they waited this long, they could be patient a little longer.

Billy, who once dreamed of game cubes and skate boards, was now thinking of the homeless and the less fortunate. He surprised himself with his own feelings. Just a few short months ago, he was a bully whose only concern was to beat up on the neighbor kid. Surprisingly, after the visit at the Goodwill Center, he could not stop thinking about the faces of the homeless people he had seen there.

Davy had planned to share some of his treasure with Mrs. Elders, and then he wanted to take his family on a nice trip, maybe the Caribbean where he imagined meeting real pirates. Davy had really been the mastermind sleuth and had finally earned the respect he so desperately wanted from his peers, especially his new friend, Billy.

Amanda wanted to give the treasure to her mom. She wanted her mom to have an easier live, working only one job instead of two. She had made some true friends here in this new town and wanted to make it a permanent home.

Pete had plans to restore the family church, since he now knew that his grandparents had so much history there. He wanted to honor his grandfather with a special deed that proved his worth as a founding family of this town. His research into his family history would allow his grandmother to move forward and possibly finish the family's bible for future generations. Just having his father home and the family together was his greatest joy. He thought of approaching his father with the idea of staying in the town of Preston and publishing his own paper since he was a reporter.

Charlie was conflicted and felt betrayed. Of all the families, his had the darkest past, and he had no history to tell him otherwise. His stepdad, a manipulating pariah, had disappeared without a trace. His mother was heartbroken and unable to function on a basic level for weeks. His grandmother's condition in the nursing home was growing worse by the day, and his younger siblings were confused. He found himself taking on more responsibility than ever. Amazingly he did not complain. He made the evening meals while his mother was visiting his grandmother, or sorting out the paper work. It became apparent that his stepdad had squandered the money that his grandmother had left them, and without the treasure, they would have been completely broke. The money from the precious stones was going to help them out.

The person who had the greatest joy of all was Pete's grandmother. Mary Fairchild James was finally able to solve the mystery of the family treasure, and share it with her own family. She was happy to know that her family fortune would be shared by her family and other fine families in her beloved town.

After all the estimates of the stones were made, every person in the small group was going to get $250,000.00. In total, that would be 1.25 million. Mary was impressed with such a fine sum of money, but the total bothered her. She knew that inflation had driven up the price of the precious stones, but even at that, the sum was lower than she had originally thought. Her father had implied that there were several millions in hidden treasure. Without letting Pete or anyone else know, Mary secretly spent her remaining days looking for more clues to the rest of the hidden treasure.